HUMAN GROWTH AND DEVELOPMENT

HUMAN GROWTH AND DEVELOPMENT

Emma O'Brien

Gill & Macmillan

Gill & Macmillan Ltd
Hume Avenue
Park West
Dublin 12
with associated companies throughout the world
www.gillmacmillan.ie

© Emma O'Brien 2008

978 07171 4461 7

Index compiled by Cover To Cover
Design and print origination in Ireland by O'K Graphic Design, Dublin

The paper used in this book is made from the wood pulp of managed forests. For every tree felled, at least one tree is planted, thereby renewing natural resources.

A CIP catalogue record for this book is available from the British Library.

Dedicated to my son Ultan
and
my grandaunt Nancy Twamley,
in memoriam

CONTENTS

ACKNOWLEDGEMENTS

While writing this book I have been fortunate to experience such generosity from others who have contributed their time and expertise. I wish to thank Professor Patricia Noonan Walsh, NDA Professor of Disability Studies, UCD, who has been a model of clarity and incisive thinking: her suggestions were invaluable to me. Elizabeth Nixon of the Children's Research Centre, TCD, and Michelle Ní Chionnaith (Savage) of Foetal Alcohol Support Ireland, both of whom were incredibly helpful. A particular note of thanks to William Creasey: his expertise in the area of physical development was invaluable and an often used resource in the development of the book. A word of gratitude to my colleagues in Inchicore College of Further Education, whose expertise and conversations have been always been stimulating, challenging and constructive. In particular Aideen Lyster and Margaret Prangnell provided feedback and suggestions for the book, as well as many cups of tea and kind words. Finally, any mistakes or omissions are mine and mine alone.

CHAPTER ONE

WHAT IS PSYCHOLOGY?

Why am I the way I am? Sound familiar? Many of us have asked the question. I came to study psychology because I was intrigued by how some people are able to overcome extreme adversity and lead successful lives yet others are not. Why the different outcomes? This question of how we become who we are is not a recent one: it has obsessed the human mind since we could reason.

History has been characterised by attempts to understand what makes us human, what shapes our thoughts and behaviour. Religion played an early part in attempting to unravel human behaviour through Christian assertion of 'original sin', the idea that man is born flawed and susceptible to undesirable behaviour.

Philosophers added to the debate as the centuries unfolded. John Locke, for instance, suggested that man was born a 'blank slate' or 'tabula rasa' and that life experiences shape who we become. Jean-Jacques Rousseau, in contrast to the Christian view, believed in the innate goodness of man striving to reach his full potential. Of course these arguments are best left to theologians and philosophers but the study of psychology is really not much different in that, put simply, it attempts to gain understanding of humans, their development and behaviours.

Psychology is the study of people; how they think, act, react and interact. Psychology is concerned with all aspects of behaviour and the thoughts, feelings and motivations underlying that behaviour. In their search for the causes of diverse forms of behaviour, psychologists take into account biological, psychological and environmental factors. (Psychology is different from psychiatry, which requires a medical degree and examines mental illness.)

THE HISTORY OF PSYCHOLOGY

Throughout the history of psychology several approaches have been used to gain a greater understanding of human behaviours, beginning in the early 1800s with:

Introspective

As the name implies, this approach relied on 'inspection', meaning that an individual would be asked to report on their feelings and thoughts. William James, who is considered one of the forefathers of psychology, was an exponent of this method, as was William Wundt.

Psychodynamic

Originating in the late 1800s this movement is best known through the work of Freud. It placed emphasis upon the 'unconscious' mind, believing that an individual has awareness of only a fraction of their thoughts and mental processes. Freud believed that unconscious urges were responsible for behaviour. Techniques such as hypnosis and dream analysis were used to access these distant recesses of the mind.

Behaviourist

This approach was very popular in America in the 1920s. Those best known for their work in this field are Skinner and Pavlov. Behaviourists believed that while the inner workings of the mind could not be observed a person's behaviour could. Their work still has some relevance in the area of learning.

Humanist

This approach, it could be argued, was a reaction against the behaviourist picture of man as almost a robot merely responding to outside influences (external stimuli) and the Freudian image of man driven by his unconscious urges. Humanists such as Carl Rogers and Abraham Maslow promoted the view that within man is an active desire to reach his full potential, or 'self-actualisation'. Their work has been important in the area of personality.

Sociocultural

No man is an island! We are social creatures. The sociocultural perspective recognises this and suggests that our thoughts and behaviours are influenced by our interactions with others. Importantly, it highlights how we are embedded in the culture in which we were raised or in which we live, and how the views of that culture in turn influence us. In the past psychology tended to look at the individual to gain a greater understanding without looking outside the person to gauge external influences on them. The sociocultural approach examines how culture is transmitted to its members and investigates the differences and similarities of people from differing cultures.

Scientific

The predominant approach in psychology at present is the scientific method, or science of behaviour. This approach is less interested in human behaviour per se, focusing instead on *why* that behaviour occurs. Thus if a child exhibits aggressive behaviour the psychologist does not focus on the behaviour itself: rather, they want to know why the child is behaving in such a fashion. Methods of research (methodology) include statistics and experiments.

Dissatisfaction has been voiced regarding the use of the scientific approach in psychology, its critics claiming that it cannot capture the complexity of human behaviour. Reactionary approaches include that of community psychology, which examines individuals within their social world. Community psychology explores social issues and how they influence individuals, groups and society at large. Nonetheless the scientific approach maintains its dominant position in psychology.

THE GOALS OF PSYCHOLOGY

1. To *describe* how people and other animals behave.
2. To *understand* the causes of these behaviour.
3. To *predict* how people and animals will behave under certain conditions.
4. To *control* behaviour through knowledge and control of its causes.

WHAT IS LIFESPAN DEVELOPMENT?

Human development is 'the scientific study of age-related changes in behaviour, thinking, emotion and personality' (Boyd & Bee 2005:3).

Lifespan development is a field of growing importance in psychology. It is the study of development across the lifespan, recognising that an individual will continue to develop and change as they move through their life. I use the word 'change' deliberately rather than the term 'growth' as some of the changes that occur across an individual's life will be those of decline, for example as we get older our physical body begins to decline.

People change and develop in many ways, and certain changes are more noticeable at different periods in a person's life. For example, the physical changes that occur in infancy are quite astounding and very obvious in comparison to the physical changes that occur in one's thirties. The study of lifespan development has become an area of growing relevance in light of the increase in life expectancy over the last hundred years.

According to WHO figures for 2005, average life expectancy at birth is 77 years for an Irish male and 81 years for an Irish female. In Zimbabwe the life expectancy for a male is 43 years and for a female it is only 42 years (www.who.int/countries/en/#Z). Even in Ireland, Traveller males have a life expectancy at birth that is ten years less than the national figures; for Traveller women the life expectancy is twelve years less (Department of Health and Children 2002b:24).

These shocking differences highlight the influence of various factors on development. These factors include:

- socio-economic status (SES)
- culture and ethnicity
- access to resources (e.g. health services).

Back to lifespan development. We have seen that it is the study of changes that occur throughout the life of an individual from conception to old age.

DOMAINS OR AREAS OF DEVELOPMENT

Physical

This area covers changes that occur in the physical body and brain. These can include height, weight, muscular and skeletal changes. Also included are sensation and perception. **Sensation** refers to the functions of the sense organs (vision, hearing, taste, touch and smell), and refers to the awareness of conditions or feelings inside and outside the body. **Perception** is how we process the information taken in by our senses. Finally, physical motor development and skills are also included. Put simply, 'motor' is another term for movement, so motor development refers to the development of movement and skills such as grasping, walking and manipulating objects.

Cognitive

At it most basic cognition refers to thinking, so cognitive development means the development of thinking or mental abilities. These mental abilities include memory, problem solving, reasoning and learning. Language is also included under cognitive development.

Social-Emotional

As already mentioned, we are social creatures, and included in this area are the relationships we have with others, our emotional and our personality development.

PATTERNS OF DEVELOPMENT

Cephalocaudal

This can be defined as development that begins at the head and travels down the body. Think of a foetus: the head develops first, and then the limbs — and more particularly the fingers and toes — are among the last areas to develop.

Proximodistal

'Proximo' relates to proximity and 'distal' refers to distance. So the proximodistal pattern refers to development from the trunk or centre of the body out to the extremities, that is the fingers and toes. If we look at a young toddler we see that they can move and control their arms before they can successfully control their fingers: this reflects the proximodistal principle at work.

It is helpful to mention 'motor skills' at this stage. Once motor development is in place (at approximately four or five years of age) the child can develop their motor skills.

A motor skill is a fundamental movement pattern performed with accuracy, precision and control. There are two types of motor skill to be aware of:

- *Gross motor skills* involve the use of large muscles to perform a movement task, for example crawling, walking and jumping. In line with the proximodistal principle, large muscles usually develop before smaller muscles.
- *Fine motor skills* involve the use of small muscles to perform a movement task with precision. This includes the manipulation of small objects, the pincer grasp (seen in young children when they grasp items between their thumb and forefinger), writing and typing.

PERIODS OF DEVELOPMENT

The lifespan is broken or categorised into phases to make study of it easier. The phases are:

- pre-natal: before birth
- infancy: 0–2
- early childhood: 2–5
- middle and late childhood: 6–11

- adolescence: 12–19
- early adulthood: 20–40
- middle adulthood: 40–65
- late adulthood: 65 onwards.

SOME KEY DEVELOPMENTAL ISSUES

Two important developmental issues are nature versus nurture and critical versus sensitive periods.

Nature vs. Nurture

The nature—nurture debate is one of the key issues in psychology and more particularly in understanding human development. Its origins can be traced to the early philosophers who discussed the nature of man. Are we 'blank slates', as Locke believed, shaped by those around us and the society we live in? Or, in keeping with the doctrine of original sin, are we born as we are?

These two positions reflect the two sides of the nature and nurture debate. **Nature** refers to biological processes, genes and our brain as determinants of our behaviour. **Nurture** relates to the influence of our environment in our development. The culture we are born into influences how we see ourselves and others, and also our behaviour. Other environmental influences would be child-rearing practices, education and so on.

In the early twentieth century the nature position dominated and this fuelled a belief in racial superiority and differences. By the 1960s this position had changed to a 'nurture' stance, which postulated the importance of environment in a person's development and can be best seen in the explosion of literature on child-rearing practices.

Nowadays a more reasonable position is generally maintained, which recognises the influence of both nature and nurture on human development. Debates continue regarding the degree to which each is involved, their role and the interaction between each element. Urie Bronfenbrenner's revised 'bioecological' theory (see Chapter 2) best captures this new position, as he demonstrates the many influences that interact to shape development. We still do not understand the complexity of the interaction between nature and nurture and it remains an issue of great interest and importance.

Critical vs. Sensitive Periods

We are going to look at the issues raised by extreme deprivation in the early years to explore the issue of critical versus sensitive periods of development. Does a critical period exist for development? In other words, if development does not occur during this time is the opportunity for development lost? Or is there a sensitive period in which it is preferable for development to occur, but if it does not, development can occur at a later date? We shall look at the issue of extreme deprivation and the debates and research that surround this issue.

In Focus: Critical vs. Sensitive Period Debate — Extreme Deprivation

Development in infancy encompasses social, emotional, cognitive and physical growth. What's the difference between critical and sensitive periods? Let's take the example of language.

If you believe in a 'critical period' for the acquisition of language you would believe that if language is not acquired during a particular period it will not be possible to learn language at a later date.

If you believe in a 'sensitive period' of development you might believe that even if you do not acquire language in the early years it is possible to learn language at a later time.

In the debate between critical and sensitive periods of development, infancy is often examined as it is a time of huge growth in many domains. The critical vs. sensitive period debate is one of the most active in psychology, and cases of extreme deprivation are often examined in an effort to illuminate this debate. Recent examples of children who have suffered extreme deprivation include the Romanian orphans whose images on television and in the press shocked us in the early 1990s. Extreme deprivation in infancy offers us an insight into whether children can recover from adverse experiences in their early years and develop normally given caring and appropriate interventions, or whether there is in fact a 'critical period' from which the individual cannot recover, leaving them permanently and irreversibly affected.

Clarke and Clarke (1999) support the idea of a sensitive period, after which experienced adversity and resulting deficits can be compensated for — an 'initial step in an ongoing life path'. On the other hand, Freud and, later, Bowlby argued that early experience determines later development. Bowlby's work with children in institutional care led him to believe that negative early experiences cannot be reversed in later years, especially in the area of attachment (Bowlby 1951).

However, in more recent decades, this position has been questioned. Cases of extreme deprivation have had quite different outcomes, suggesting that the debate surrounding critical vs. sensitive periods isn't quite as clear-cut as earlier presumed.

Rutter argues that 'even markedly adverse experiences in infancy carry few risks for later development if the subsequent rearing environment is a good one' (1989:24).

Let's consider how true this is by examining some cases of extreme deprivation and their subsequent outcomes.

Skuse (1984a) explored case studies of children who had spent their early years in conditions of extreme adversity and deprivation, hoping to explore specific questions.

- Are some psychological qualities more sensitive to deprivation than others?
- At what pace does recovery take place and what course does it follow?
- What interventions are necessary to optimise recovery?

> Anna was discovered in 1938 at nearly six years of age, having spent her life in a storage room tied to a chair with her arms above her head. Severely malnourished, she was skeletal, expressionless, lacking speech, with severe motor retardation. While Anna showed some improvement, she never integrated successfully into her peer group, even though she was now living with a foster family. Anna received no specialist intervention and while she made some improvement she remained severely retarded until her death at the age of ten. (Clarke & Clarke 1976:28).

> Isabelle was discovered at six years old locked in a dark room with her 'deaf-mute' mother. Isabelle was suffering from severe malnourishment, her behaviour was either infantile or like that of a wild animal and she did not seem to possess speech. Experts decided she was 'feeble-minded' (Clarke & Clarke 1976:42). Within two years of intensive speech and educational therapy she had achieved a normal level of speech and cognitive function.

Language

Let's examine a specific area of development, that of language, to see if we can come closer to a conclusion in this debate of critical vs. sensitive periods.

Skuse concludes that language appears to be the most vulnerable to deprivation, but much debate surrounds the question of whether there is a critical period for language exposure and/or acquisiton. Hall replied to Skuse with the suggestion that 'some exposure to language and communication is essential at a very early stage, even if only for a very brief period' (1985:825). While Lenneberg's 'critical age hypothesis' (cited in Curtiss 1977:208), states that the critical period runs from two years to puberty.

Genie was confined in total isolation from 20 months until her discovery at 13 years of age; although intensive language therapy suggested initial promising acquisition, her capacity remained severely limited and she was capable of 'few normal or appropriate acts of communication at 18 years' (Skuse 1984b:562).

Meanwhile Isabelle, who was confined with a deaf-mute mother, developed normal language skills.

Or we could consider the case of the Koluchova twins, grossly deprived and confined from 18 months to nearly seven years, who went on to develop normally with respect to language (as well as in every other facet of life) (Skuse 1984b). Was this due to their early normal language exposure before they were confined? Was it because they had each other for company and developed ways of communicating? Or was it due to their later intensive care within a foster family?

English and Romanian Adoptees (ERA)

We will now turn to the work of the English and Romanian Adoptees (ERA) study team and their ongoing research comparing UK adoptees with those adopted into the UK from Romania. The fall of the Ceausescu regime in Romania created a unique research opportunity following the humanitarian endeavour of removing infants and children from orphanages in which they had suffered severe deprivation.

Study

The research (Rutter & ERA 1998, 2000) looked at 324 children who were adopted into the UK.

The conditions in which the children had lived in Romania varied from 'poor to appalling' (Rutter & ERA 1998:467). They were confined to cots; had few, if any, playthings and barely any stimulation through talk or play; were poorly nourished; and endured often harsh physical conditions. Nearly half had been reared in institutions throughout their lives; 18 of 111 had had family rearing with only two weeks' institutional care; the rest had spent about half their lives in institutions. Half were severely malnourished and suffering from chronic infections.

Findings

- The catch-up with respect to these norms was nearly complete at age four (Rutter & ERA 1998).
- Age on adoption was a strong predictor of more positive outcomes with no measurable deficit in those adopted pre-six months. Those adopted post-six months

were more likely to show evidence of deficits.

- Children who had received better individualised care in institutions tended to have higher cognitive scores at age six (Rutter *et al*. 2000).

Rutter *et al*. conclude that 'children who had experienced prolonged privation in poor quality institutions tended to show a less complete cognitive recovery, although even with prolonged institutional care, cognitive catch-up was very substantial indeed' (2000:111).

Irish Findings

In 2007 Dr Sheila Greene of the Children's Research Centre, TCD presented a paper on 'Children's recovery after early adversity: lessons from inter-country adoption'. The findings were based on research conducted by the Children's Research Centre on inter-country adoption in Ireland. In the presentation Dr Greene concluded:

- Inter-country adoption provides many striking examples of resilience.
- Pervasive environmental change from very poor to very positive circumstances can bring about remarkable levels of recovery in children suffering the effects of adversity –– by any standard.
- Most children demonstrate a capacity to recover when their circumstances change dramatically; a minority do not.
- Some children who have been subjected to long periods of very intense deprivation will show some recovery but may never be normal in their functioning (p.14).

Further information on the research into inter-country adoption can be accessed at: www.tcd.ie/childrensresearchcentre/.

Our understanding has certainly extended far beyond early simplistic suggestions that early experience determines all future development. What can cause these differences in outcomes?

Some no doubt are accounted for by biological and genetic factors which are difficult to pinpoint in each case.

Other factors include temperament, personality and environmental variation. What is apparent is that there are no clear-cut answers to this complex developmental issue.

CHAPTER TWO

THEORISTS

T heories have been proposed, some newer than others, trying to explain human development or aspects of it. They are generally constructed in an attempt to explain and predict behaviour. Some of these theories relate to specific areas of development such as cognitive development (Piaget), while others are influenced by a particular approach in psychology, such as the psychoanalytic movement (Freud, Erikson).

The main theorists we will look at are:

- social and emotional development: Sigmund Freud (1856–1939); Erik Erikson (1902–94)
- cognitive development: Jean Piaget (1896–1980); Lev Vygotsky (1896–1934)
- ecological: Urie Bronfenbrenner (1917–2005); Glen H. Elder Jr (1934–)

SOCIAL AND EMOTIONAL THEORISTS

Social and emotional theories relate to the development of our emotional selves and to personality development. Social development refers not just to our relationships with others and how we construe them; it also involves how we see ourselves in the eyes or from the perspective of others. An example would be **social comparison,** which means comparing ourselves to another person. If you've ever looked at a magazine with an image of an impossibly thin woman and felt the urge to start a diet, that's an example of social comparison. So as you can see social development has two aspects. The two theorists that we are going to examine come from the psychoanalytic tradition, yet they have quite different explanations of this aspect of human development.

Sigmund Freud (1856–1939)

Sigmund Freud is considered the founder of the psychoanalytic movement. What is psychoanalysis and how does it relate to human development? As mentioned in the previous chapter, Freud believed that much of human behaviour, emotions and feelings emanated from the unconscious mind. Our behaviour is merely the tip of the iceberg that is the unconscious mind. In order to understand behaviour Freud believed that the unconscious mind would need to be accessed through techniques such as word association and dream analysis. We are not aware of this part of our mind, yet Freud believed that it holds powerful urges that had to be satisfied. This creates the impetus for our behaviour.

Conscious/Pre-Conscious/Unconscious Mind

Benson (2003:49) provides the following explanation.

The **conscious** (the top one seventh) is the awareness we have when we are awake.

The **pre–conscious** (the boundary) contains memories of dreams, slips of the tongue, etc. It gives clues about the unconscious in the thoughts and actions that appear there. If you have a dream, you are not directly revealing unconscious thoughts but recalling highly coded ideas. This symbolism protects us, so that we are not upset or disturbed by what our unconscious is really thinking.

The **unconscious** (six sevenths of our mind) contains secret wishes and fears, traumatic memories of the past, etc. All these thoughts are completely hidden and totally unavailable to us. This is necessary for survival — we have to forget past traumas in order to get on with our lives. We can **never** see directly into the unconscious.

Components of Personality

According to Freud the personality comprises three components: the id, ego and superego.

The **id** is the only structure present at birth and it exists solely in the unconscious mind. Freud (1964:106) describes the id as 'a chaos, a cauldron of seething excitations' and further says that 'it is filled with energy reaching it from the instincts, but it has no organization, produces no collective will, but only a striving to bring about the satisfaction of the instinctual pleasure principle'. The pleasure principle is solely concerned with instant gratification regardless of any other consideration. In order to control the impulses of the id another aspect of personality develops, the ego.

The **ego** operates primarily at a conscious level and is driven by the reality principle, which acts as a check on the id. The ego examines whether it is safe for the id to discharge its impulses and satisfy its instinctual needs.

Finally the last component comes into play: the **superego**, which in its barest form is the moral voice of the personality, endeavours to control the id, particularly with regard to sexual and aggressive needs.

The superego grows out of a child's internalisation of their parents' views of morality and is influenced by the cultural norms of the society of which the individual is a member. The ego stands in the middle between the id and the superego.

Think of it another way. As we get older we are taught by parents, school and our culture that expectations of us exist and that our needs cannot always be satisfied. For example, in many cultures stealing is prohibited and frowned upon. I'm hungry and go into a shop to buy some food when I realise I've forgotten my purse. An 'id' response would be to take and eat the food — the id is only interested in satisfying my hunger. However, I've been brought up to believe that stealing is wrong, so I put the food back — this is a 'superego' response. The superego develops as we get older and learn more and more of what is acceptable and the rules we must obey (legal and moral).

Of course the id and the superego both represent extremes, and that's where the 'ego' comes in. Like the superego it develops as we get older and it acts as a kind of mediator between the id and the superego, trying to find an outcome that will satisfy both. To go back to the shop example, after I've discovered I've forgotten my money the 'ego' solution could be for me to ask my friend to lend me the money so I can buy the food. That way my hunger will be satisfied, keeping the id happy, and I won't have stolen anything, keeping my superego happy — and also my mother!

As can be imagined, conflict occurs between the competing demands of the id for satisfaction at any cost and the superego attempting to constrain the id. **Anxiety** occurs when the ego confronts impulses that are threatening to get out of control or external danger. According to Hall (1962) anxiety is a painful emotional experience. When the ego is unable to produce realistic coping mechanisms in times of anxiety it resorts to what Freud termed 'defence mechanism' to relieve the anxiety in a safe manner by denying, distorting or falsifying reality. A number of defence mechanisms exist, including repression, projection, reaction formation, denial, displacement and regression. The purpose of defence mechanisms is a developmental one: they appear early on and are used by the infantile ego to cope with demands and as a protective measure. Freud claimed that defence mechanisms continue after the infantile stage if the ego has not developed sufficiently and has remained dependent on the defence mechanisms.

Defence Mechanisms

Repression is one of the most widely known of the defence mechanisms and was of primary interest to Freud. In the beginning Freud coined the term 'repression' to describe unacceptable or painful mental contents excluded from awareness. These drives and memories trapped in the unconscious strive for release and occasionally will escape in the form of a slip of the tongue, a 'Freudian slip'. An example of repression is when an individual who was abused as a child develops amnesia of the event.

Denial is another well-known defence mechanism. It involves the denial of an event (or emotions associated with an event) that is painful and generating high levels of anxiety. While in repression the individual is rendered unaware of an event, in denial the person is aware but refuses to acknowledge the reality. They deny it in an attempt to avoid acknowledging the extent of the threat. An alcoholic, when challenged about their drinking, will deny that they have a problem.

Freud and his supporters contend that denial through a refusal to acknowledge the reality of a situation is necessarily a bad thing as it impedes good mental health; they propose that 'reality orientation' is essential to mental health.

Projection is considered another primitive defence mechanism. It occurs when an unacceptable impulse is repressed and turned or 'projected' outwards and attributed externally. Thus an individual with internal hostile impulses represses this and instead views others around them (external) as hostile. Projection is an inability to recognise hostility in oneself and instead attributing the characteristic to others. The old saying 'a liar never believes anyone else' captures the essence of projection: an individual with a propensity to lie projects this characteristic onto others, thus believing themselves to be honest and those around them to be liars.

> DISCUSSION TOPIC
> What do you think of defence mechanisms?

Freud's Psychosexual Theory

Freud proposed a 'stage theory': an individual must pass through one stage to reach the next stage of development. As we will see, Freud believed that each stage could have a negative outcome and the individual could become 'fixated' or stuck at that stage.

Importantly, Freud was arguably one of the first lifespan developmentalists since he believed that these early experiences could be responsible for behaviour or personality traits in later life. Freud stressed the importance of early experiences in determining the outcomes of later life for an individual.

Freud describes five stages of psychosexual development, which we all pass through:

- oral (0–2 years)
- anal (2–3 years)
- phallic (3–6 years)
- latent (6–11 years)
- genital (11+ years).

Benson (2003:53–5) takes us through the first three stages.

Oral stage. The mouth is the prime source of pleasure: for survival, the baby instinctively sucks. Through oral satisfaction, the baby develops trust and an optimistic personality. Being stuck at this stage is 'oral fixation', for example if the baby is weaned too early the personality may become pessimistic, aggressive and distrustful.

Anal stage. The focus of pleasure shifts to the anus, helping the child become aware of its bowels and how to control them. By deciding itself, the child takes an important step of independence, developing confidence and a sense of when to 'give things up'. However, over-strictness about forcing a child to go or about timing or cleanliness can cause personality problems. In anal fixation, forcing a child to go may cause reluctance about giving away *anything*. The person may become a hoarder or a miser. Conversely, over-concern about 'going regularly' may lead to obsessive timekeeping or always being late.

Phallic stage. Children become aware of their genitals and sexual differences. Consequently, development is different for boys and girls.

The Oedipus Complex: each boy unconsciously goes through a sequence of stages beginning with the development of a strong desire for his mother, noticing the bond between his parents (i.e. sleeping together), becoming jealous of his father and hating him, then becoming afraid of his father lest he discover his son's true feelings, which results in the final substage of fearing punishment — which is castration!

Latency stage. During this stage sexual urges remain repressed and children interact and play mostly with same-sex peers.

Genital stage. The final stage of psychosexual development begins at the start of puberty when sexual urges are once again awakened. Adolescents direct their sexual urges towards opposite-sex peers, and the primary focus of pleasure is the genitals.

Criticisms of Freud's Psychosexual Theory

- Freud's theory is difficult to test and results from the few tests carried out have not been favourable.
- Some of the concepts Freud discusses are difficult to test as they cannot be observed.
- Freud's theory of 'penis envy' has been attacked by some feminists as sexist.
- His observations were of adults only. He did not conduct research on children in developing his theory of psychosexual development.

Erik Erikson (1902–94)

Erik Erikson was born in Germany and became interested in psychology when he met Anna Freud (Sigmund Freud's daughter), who convinced him to study child psychoanalysis in Vienna. While Erikson was influenced by Freud and accepted some of what he said, he believed that Freud was incorrect in his proposition of psychosexual stages, instead proposing that individuals develop through psychosocial stages. Erikson moved to America and joined Harvard Medical School before moving to Yale.

Erikson's first book, *Childhood and Society*, reflected his interest in the role of society and culture in the development of the child. Erik Erikson suggested that personality develops through the resolution of a series of eight major psychosocial stages occurring throughout the individual's life. Each stage involves a different 'crisis' or conflict between the 'self' (individual) and others, including the outside world, which can result in either a positive or negative outcome. For example, in his first stage — 'trust vs. mistrust' — if the infant is well cared for and its needs met it will develop trust (a positive outcome of this stage). However, if the infant is mistreated or abused a negative outcome of 'mistrust' will occur. Erikson was more positive than Freud as he believed that negative outcomes in a stage could be resolved at a later date.

The table opposite outlines each of Erikson's eight psychosocial stages.

Erikson was one of the first theorists to see development in a lifespan context. His eight psychosocial stages, called 'The Eight Ages of Man', cover the entire lifespan of the individual.

Trust vs. Mistrust (birth to 1 year). In the first year of life the infant is completely dependent on its caregivers. How well the infant's needs are met and how sensitive the parenting is will decide whether the infant develops trust or mistrust of the world.

Autonomy vs. Shame and Doubt (1 to 2 years). Children begin to walk and assert their independence. As the term 'autonomy' suggests, children can come to believe in themselves and their abilities through the encouragement and support of their caregivers. If the child's efforts are ridiculed or belittled the child will develop a feeling of shame and doubt about their abilities. Toilet training is seen as a key event that can influence the outcome of this stage.

Initiative vs. Guilt (3 to 5 years). As the child becomes older they exhibit increasing curiosity and interest, they initiate play and question more. If this initiative is discouraged or they are held back they will not develop self-initiative and instead will hold back in later life.

Table 2.1: Erickson's eight psychological stages

Approximate Age Period	Stage	Psychological Realionship
Birth–1 year	Trust vs. Mistrust	Develops trust in others and the world or Suspicion and mistrust
1–2 years	Autonomy vs. Shame and Doubt	Sense of self-reliance or Feelings of shame about one's capability
3–5 years	Initiative vs. Guilt	Ability to start activities or Guilt about feeelings
6–12 years	Industry vs. Inferiority	Sense of confidence in ability to do things or Feelings of inferiority based on reactions of others
12–20 years	Indentity vs. Role Confusion	Develop sense of who you are or Confusion as to who you are and role in life
20–40 years	Intimacy vs. Isolation	Experience love and form relationships or Isolation, shallow relationships
40–65 years	Generativity vs. Stagnation	Seek to be productive or Lack of growth and boredom
65 years onwards	Integrity vs. Despair	Satisfaction with one's life or Regret over missed opportunities

Industry vs. Inferiority (6 to 12 years). During this time the child begins to attend school and they interact more with peers. If praised in their efforts they develop a sense of industry and feel good about what they have achieved, which encourages the feeling that they can fulfil their goals. If they repeatedly fail or are not praised when they try, a sense of inferiority will develop.

Identity vs. Role Confusion (12 to 20 years). This relates to the adolescent period of life when the teenager is establishing their identity, their sense of who they are and their role in life. They are becoming more independent and their peers are increasingly important to them. If they can reconcile these issues they will develop a feeling of identity; if they cannot, role confusion will occur.

Intimacy vs. Isolation (20 to 40 years). This period is marked by the desire to establish relationships with others. Successful completion leads to a sense of security and intimacy. Avoidance of intimacy or an inability to establish a secure relationship leads to feelings of isolation.

Generativity vs. Stagnation (40 to 65 years). During this period we settle down in a relationship and perhaps bring up children. People also establish their careers. A sense of community and being part of a bigger picture becomes important. If we do not achieve these objectives the resulting feeling is one of stagnation.

Integrity vs. Despair (65 years onwards). In older age, people slow down and begin to reflect on their lives. If the individual feels they have had a successful life, a sense of integrity prevails. If they believe their life to have been unproductive a feeling of despair occurs.

COGNITIVE THEORISTS

Cognitive development relates to the changes that occur in thinking, memory and problem solving, to name a few of the mental abilities that are considered cognitive. Jean Piaget was one of the most famous and important theorists in the field of cognitive development, and his work is still influential in areas of psychology, education and child development.

Jean Piaget (1896–1980)

Jean Piaget was born in Switzerland in 1896. After completing his PhD, he developed an interest in psychoanalysis and travelled to France to work in a boys' institution which had been founded by Albert Binet. Binet is best known for his work in developing intelligence tests and Piaget worked on standardising these tests. It was through his work on the tests that Piaget came to the conclusion that young children think differently from adults. He noticed that many of the children were giving the same 'wrong' answer. Piaget began to understand that children handle information differently from adults, though as the child grows older their thinking becomes more like an adult's.

These observations prompted Piaget to develop a theory of cognitive development.

His theory is a 'stage theory': the individual must pass through one stage before they can progress to the next. He suggested that children pass through four stages of cognitive development spanning infancy to adolescence. Interestingly, Piaget was a forerunner of 'naturalistic observation': this approach to collecting data relies on observing individuals in their natural environment, rather than in a laboratory. Piaget and his wife Valentine observed their three children from infancy and kept detailed journals noting their intellectual development. Piaget can be daunting and difficult to understand, especially his terminology, but don't let that discourage you. What Piaget is trying to do is outline how he believes children organise the information that they receive in their daily lives.

Before we look at the stages of his theory it's helpful to understand how Piaget believed learning occurs. **Schemas** is the term he used to describe 'internal frameworks' that the mind builds as it comes into contact with more and more information. It's the equivalent of building structures that can hold and make sense of incoming information. As the infant gathers more and more information the schemas become more sophisticated.

Definition schema is an internal framework that organises incoming information, thought and action.

Piaget believed that in order for learning to occur the child must experience a sense of 'disequilibration'. This is when an experience occurs that does not fit the child's existing thinking: the individual becomes dissatisfied with their original thinking and a state of 'disequilibration' occurs. Because it won't fit, the child must adapt in order to process the new piece of information.

This can be done in two ways: adaptation through **assimilation** and **accommodation**.

Assimilation is when the person fits incoming information or experience into an existing schema. For example, a young child who sees a fox for the first time might call it a 'doggie'. Here the child is trying to make sense of this new experience by applying an existing 'doggie' schema to the fox, which has four legs and a tail. This make sense as a dog also has four legs and a tail. Yet at we get older we come to understand that a fox and a dog are not the same and this might cause us to change our schema. This process is called **accommodation**.

Definitions: accommodation is when an existing schema changes to incorporate new experiences; assimilation is when the individual fits incoming information or experience into an existing schema.

Table 2.2: Piaget's theory of cognitive development

Age	Stage	Characteristics
0–2 years	Sensori-motor	Knows about the world through movement and sensory information Learns to differentiate self from environment Develops first schemas Achieves object permanence Symbolic thought emerges Capacity to form internal mental representations emerges
2–7 years	Pre-operational	Symbolic thinking emerges as child uses symbols/images to represent objects and problem solve Begins to understand classification of objects Egocentric Focuses on just one aspect of a task Animism (believes inanimate objects have consciousness)
7–11 years	Concrete operational	Undertsands conservation of mass, lenght, weight and volume Decentring – becomes less egocentric, taking more easily the perspective of others Reversability also a feature Can classify and order, as well as organise objects into series Still tied to the immediate experience but within these limitations can perform logical mental operations
12 years onwards	Formal operational	Abstract thinking marks this period – the individual is now able to manipulate ideas in their head Inductive and deductive reasoning emerge, enabling the formulation and testing of hypotheses

Adapted from Cowie & Smith, *Understanding Children's Development* (3rd edn), p. 336

Sensori-motor Stage (0–2 Years)

'Sensori' represents the senses or sensation (taste, touch, smell, vision and hearing), 'motor' is another term for movement. Thus the name gives us a clue as to how learning is accumulated during this period — through the senses and movement. Have you ever noticed how young babies are forever putting things in their mouth? This is because they are using this sense to explore the new world they inhabit. Also, as the infant begins to crawl they usually head straight for the kitchen presses, much to their parents' consternation, which again reflects the child's attempts to use their new-found movement to explore their environment further. The infant is hungry for new experiences and information that will form the basis of their developing schemas.

By the end of this stage infants will have acquired 'object permanence'. This is a really important concept and represents the understanding that objects continue to exist even when we can no longer see them. A simple test for object permanence is to allow the child see a toy and then throw a blanket over it. If the child seeks the toy out it is an indication that they understand that the toy continues to exist even when they could not see it. This ability comes towards the end of this stage.

Pre-operational Stage (2–7 Years)

One of the most notable features of this stage is that the child is **egocentric**. If you break the word up it gives a clue to its meaning: 'ego' means self and 'centric' refers to the centre of things. So, in effect, egocentric means self-centred. However, when used in this context, it refers to the fact that the child thinks that everyone else sees the world through their eyes. They do not understand that other people might see or think about things differently, and this is reflected in their thinking.

Symbolic thinking appears as the child begins to acquire language (which is of course made up of symbols). They can use this newly developed ability to enhance their thinking. The child can now use words and images (symbols) to represent objects, for example, when a toddler sees a dog they might exclaim 'woof woof', using the noise to represent a dog. *Pretend play* is another beneficiary of symbolic thinking — now a stick is transformed into a sword, for example. An interesting aspect of this stage is *animism*. Have you ever smacked a bold table that a young child has just run into? Children at this age ascribe consciousness to inanimate objects, so the chair now becomes a naughty, bold chair that is responsible for that knock to the head!

Concrete Operational Stage (7–11 Years)

Children learn through concrete or 'real' objects. Piaget believed that the acquisition of *conservation* was an important developmental milestone (see diagram on p. 97). There are many types of conservation (length, mass and weight), but we'll look at conservation of liquid. Have you ever been at a party when a row has broken out between children because one child has more juice than another? As much as you try to explain that although the glasses are different shapes the amount of liquid in each is the same the child refuses to accept this. This is because the child lacks conservation. Another goal that is reached at the conclusion of this phase is *reversibility*, which is needed to enable the child to grasp conservation. Reversibility is the ability to mentally undo an action.

Decentring occurs during this period. The child becomes less egocentric in their thinking and able to take the perspective of others.

Formal Operational (12 Years Onwards)

This final stage is marked by the acquisition of *abstract thinking*. The individual is now able to manipulate ideas in their head. Up to this stage the child can only think about things that are 'real' or concrete. According to Piaget they are unable to reason about make-believe problems or situations. Now they can begin to think more logically and they can consider hypotheses or explanations. Inductive and deductive reasoning emerges during this stage, enabling further formulation and testing of hypotheses.

Piaget and Play

Piaget conceived that the development of play was broken into three parts.

Play Stage	Age	Piagetian Stage	Type of Play
Mastery Stage	1–2 years	Sensorimotor	Plays alone. Features of play include repetition and mastering behaviours through imitation.
Play Stage	3–6 years	Preoperational	In the early years children are too egocentric to play with others: play is either alone or parallel. Use of symbolic thinking in play, objects and actions used symbolically. Egocentrism begins to ease towards the end of this stage.
Game Stage	7 years	Concrete operational	No longer egocentric, children begin to play together, especially in a co-operative manner. Language is used in establishing rules and roles for games.

Lev Vygotsky (1896–1934)
Sociocultural Theory

Vygotsky was Russian and was born in the same year as Piaget, yet it is only in recent years that his writings have come to the attention of the West. Vygotsky died at a young age and in the years preceding his death he had been under pressure from the government to modify his beliefs and teaching in line with the current orthodoxy. After his death his theories and ideas were repudiated by the Russian government and it is only because his students kept his work alive that we in the West have come to know his work.

Vygotsky is a cognitive theorist but he is sometimes described as a sociocultural theorist. This is because he emphasised the role of others in the development of learning. Unlike Piaget, Vygotsky believed that social interactions were of particular influence in the learning of a child, as was their wider society, and it is this emphasis that has led to him been described as a sociocultural theorist.

The main points in his thinking are:

- children learn from others (including other children)
- the importance of play in the development of learning
- language plays a central role in mental development
- language and development build on each other
- development cannot be separated from its social context.

Importance of Play

Like Piaget, Vygotsky emphasised the importance of play. Piaget emphasised the child as a solitary learner: when they play they discover new ideas for themselves. Vygotsky, on the other hand, believed that children learn through their interactions with others who introduce them to new concepts and ideas.

Zone of Proximal Development

The zone of proximal development represents the distance between what the child can actually achieve on their own and what they could achieve with the intervention or help of another. Vygotsky was interested in the role others can play as a 'scaffold', building a bridge to help the child reach their full potential development. For example, as an adult I started going to Irish classes because I was not as fluent in Irish as I wished to be. My tutor ('other') observed the level of Irish I did have and recognised what I could achieve with her help, so she devised a learning plan to enable me to reach my potential. The trick for the tutor was to make sure that it was challenging enough while ensuring it was not so far beyond my ability that I could not do it, even with her help.

As we've seen, Vygotsky felt that it was through interaction with others (teachers or peers) that children learn. This applies to an adult trying to brush up on their Irish as much as to a young child learning to tie their shoelaces.

Relationship between Language and Thought

Vygotsky believed that language represents an opportunity for social interaction and learning. Further, the shared experience that language brings is necessary for the development of cognitive ability. When children begin to talk it opens a window into their minds and we can begin to understand their thought processes.

Cultural Context

It is through the child's interaction with others that they learn the culture they are part of, including language and belief systems.

ECOLOGICAL THEORIES

Urie Bronfenbrenner (1917–2005)

The ecological theory looks at a child's development within the context of the system of relationships that form his or her environment. You can see from Bronfenbrenner's theoretical model that there are several systems that impact on the individual child's development. The really clever aspect of this approach is that it encompasses the immediate environment of the child (parents, siblings) and more distant influences at the outer circle (social welfare payments for example). We're going to look at each of the 'circles' in the child's environment.

Microsystem

This circle takes in the immediate environment of the child and includes family, school/teacher, peers and neighbourhood. Thus anything that the child interacts directly with or has a relationship with can be included in this circle. Not only do these influences impact on the child, the child also impacts on its immediate environment. Relationships at this level impact in two ways: Bronfenbrenner would call this a bi-directional influence.

Mesosystem

This can be a little trickier to understand. Essentially it refers to connections or relationships between different microsystems (for example, linkages between home and school). If a child has a difficult relationship with their parents this can influence their interactions with peers (linkage between two microsystems, in this case parents and peers).

Exosystem

While the child does not have an active role and is not in immediate contact with it, this system can nonetheless affect what the child experiences in its immediate context. For example, work problems can affect the relationship between parents and also between parents and child. If the father is promoted in work this might mean he spends less time at home, leading to arguments with his wife and less time spent with his child. So the

Figure 2.1: Bronfenbrenner's circles of environment

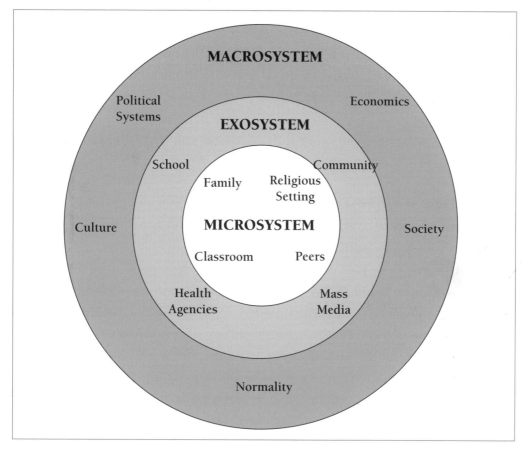

child is inadvertently impacted upon by his/her father's workplace. The educational system, government agencies and mass media are influences that are seen in the exosystem.

Macrosystem

This system refers to the culture and belief patterns of the people/society in which the child lives.

Chronosystem

This relates to changing sociohistorical circumstances. Put simply, each generation has different experiences from earlier generations. In the 1950s in Ireland, a female civil servant had to leave her job when she married; in 2008 women can marry and continue to work. This is just one change that has occurred with the passage of time: there are many others.

Initially Bronfenbrenner's theory was a strongly ecological one, emphasising the effects of nurture or environmental influences. He later modified his theory to include the biological aspect of the child, reflecting the modern view that a child's biology and environment both interact in the child's development. This modified theory is referred to as Bronfenbrenner's bioecological theory. If you look at the model you will see that aspects such as the child's sex, age and health are included as factors that influence development.

So, given that nature continues on a given path, how does the world that surrounds the child help or hinder continued development?

Glen H. Elder Jr (1934–)

Life-course Theory

Elder's theory represents another approach to conceptualising and understanding human development across the life course. Elder stresses the influence of social forces on shaping the life course and its developmental consequences. According to Elder, life-course theory represents a general change in how we think about and study human lives and development. Where Bronfenbrenner's ecological theory now encompasses the biological influence of the individual, for Elder the life course has evolved into an effective way of investigating the impact of social change on the developmental course of human lives. Instead of concentrating on individual case studies, Elder's attention focuses on multiple and interlocking pathways. Let's take a closer look at how Elder constructed his theory. Four themes are distinctive:

1. human lives in historical time and place
2. human agency and social constraints
3. the timing of lives
4. linked lives.

Human Lives in Historical Time and Place

Elder conducted research looking at the lives of boys who grew up during the Depression in America and those who grew up in Manchester. It showed that they had very different life chances following their involvement in World War Two. 'The California boys managed to escape the limitations of their deprived households by joining the armed forces and, after the war, using the benefits of the GI Bill for higher education' (2001:37). We can see how historical events and place can affect the lives of individuals.

Human Agency and Social Constraint

'Within the constraints of their world, people often plan and choose within options that become the building blocks of their evolving lives ... people of the same age do not march in concert across the major events of their life course; rather, they vary in pace and sequencing, and this variation has real consequences for people and society' (2001:38). Having children, starting work or getting married (if at all) occur at different ages under differing circumstances that reflect the individual's personal life experiences and their interpretation of a given situation.

The Timing of Lives

Elder explains that 'social timing refers to the initiation of and departure from social roles, and to relevant age expectations and beliefs. The social meanings of age give structure to the life course through age norms and sanctions, social timetables for the occurrence and order of events, generalized age grades (such as childhood and adolescence), and age hierarchies in organizational settings (i.e., the age structure of firms)' (2001:38).

Linked Lives

Most people's lives are intertwined with others, whether immediate family or more distantly placed work colleagues. Another interesting slant is the intergenerational aspect to this theme of linked lives. A failed marriage can impact on the children when as adults their life experience can be linked to the misfortunes of their parents, i.e. they may be at increased risk of marital breakdown themselves. Elder comments, 'each generation is bound to fateful decisions and events in the other's life course' (2001:39).

Finally, another component in Elder's life-course theory is the 'cohort effect', which refers to '... one of the ways in which lives can be influenced by social change. History is experienced as a cohort effect when social change and culture differentiate the life patterns of successive cohorts' (2001:37).

What does this mean? Let's take the example of a child born with Down Syndrome in Ireland. Compare and contrast the likely experience of that child if they had been born in the 1940s and if they had been born in 2000.

The following is an extract from Griffin and Shevlin's *Responding to Special Educational Needs: An Irish Perspective*. It highlights the attitudes and experience of those born with a disability in the 1940s and refers to Dr Louis Clifford's 1943 survey of learning disability.

Having a disabled child, according to Clifford, was widely seen by parents of the time as a disgrace and a reflection on the family. The more affluent and socially superior the family, the more the condition was resented and abhorred. Families from lower socio-economic backgrounds were usually more philosophical about their misfortune. In his account, Dr Clifford records that disabled children were sometimes hidden away in top rooms and seldom taken out except at night (Griffin & Shevlin 2007:39).

From this perspective, it's hard to have imagined that children and adults would be participating in an event such as the Special Olympics (the global Special Olympics began in 1968), yet change does occur.

Does a child born today with Down Syndrome face a different experience than they would have done if they were born in the 1940s?

If your answer is 'yes', this represents the 'cohort effect' — the life experience of a generation differs from that of another because of social changes. It also demonstrates the power and influence that social change can produce. This is particularly important in areas where people, such as those with disabilities, experience prejudice as it demonstrates the power of social change and the difference it can make to the lives of individuals.

In Focus: National Children's Office — Whole Child Perspective
The 2000 National Children's Strategy is a ten-year plan with a vision of:

> An Ireland where children are respected as young citizens with a valued contribution to make and a voice of their own; where all children are cherished and supported by family and the wider society; where they enjoy a fulfilling childhood and realise their potential. (Department of Health and Children 2000b)

The three national goals of the strategy are:
1   Children will have a voice in matters which affect them and their views will be given due weight in accordance with their age and maturity.
2   Children's lives will be better understood; their lives will benefit from evaluation, research and information on their needs, rights and the effectiveness of services.
3   Children will receive quality supports and services to promote all aspects of their development.

Figure 2.2: The 'whole child' perspective

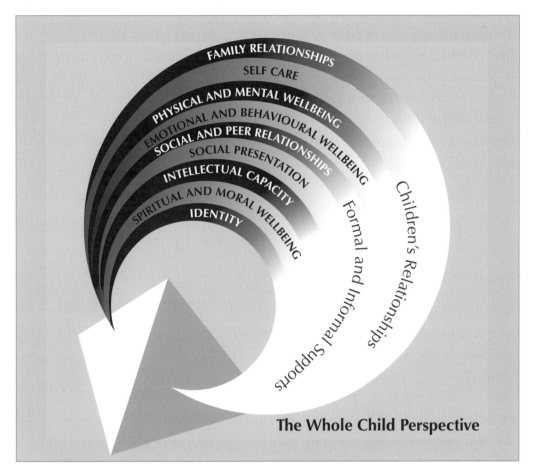

It includes a range of actions in such areas as: giving children a voice so that their views are considered in relation to matters that affect them; eliminating child poverty; ensuring children have access to play and recreation facilities; and improving research on children's lives in Ireland. The strategy provides the first comprehensive national policy document for the full range of statutory and non-statutory providers in the development of services for children and is underpinned by the United Nations Convention on the Rights of the Child.

The Whole Child Perspective
The strategy adopts a 'whole child perspective', recognising the multidimensional nature of all aspects of children's lives. The recognition that all parts of children's lives are interlinked has, in turn, implications for public policy-making and the integration of services relating to children.

What is interesting about this 'whole child perspective' is that it takes an ecological approach in attempting to understand children's development. Bronfenbrenner, in his model, emphasises the importance of recognising that systems of influence exist in a child's life from immediate family to more distant influences, such as government policy. In looking at the model of the whole child perspective developed by the National Children's Office we can see the many and varied influences they include in order to gain understanding of the child and the context of the child. So we can see evidence of a theoretical approach being adopted by a government agency whose aim is to improve the lives of children.

CHAPTER THREE

CONCEPTION, PRENATAL DEVELOPMENT AND BIRTH

CONCEPTION

Often when people think of human development across the lifespan the starting point is birth. However, many of the foundations for development occur at the moment of conception. This chapter will begin by exploring the processes that occur at conception and the possible impact that they can have on a person's development after birth.

Fertilisation/Conception

The menstrual cycle occurs in females of reproductive age. At the beginning of the cycle an egg (*ovum*) begins to develop in the ovary. This egg is released and travels down the Fallopian tube. *Fertilisation* occurs when a women's ovum or egg is penetrated or fertilised by a man's sperm. The fertilised egg is called a *zygote*. A continuous process of cell division then occurs until this mass of cells attaches itself to the mother's uterus, approximately 10–14 days after fertilisation occurred. The man's sperm and the woman's egg each contain hereditary information in the form of *chromosomes*.

Chromosomes and Genes

A *chromosome* is a tightly coiled molecule of deoxyribonucleic acid (DNA) that consists of smaller segments called *genes*. Genes are rather like an instruction manual for the body: they affect your looks, your health and how your body works. The zygote contains 23 pairs of chromosomes: 23 single chromosomes from the sperm; and 23 chromosomes from the egg (in total 46 chromosomes/23 pairs). Thus, in the fertilised egg each pair of chromosomes contains one chromosome from each parent. Out of the 23 pairs of chromosomes, it is the final pair (the 23rd pair) that determine a person's sex:

• in males the 23rd pair consists of an X and a Y chromosome

• in females the 23rd pair consists of two X chromosomes.

There are approximately 25,000 genes contained on the 46 chromosomes in each cell in a human body. This means that one chromosome contains thousands of genes. We inherit our genetic make-up from our parents through the transmission of their genes during conception.

Genotype is the sum total of our genetic make-up, the specific genetic make-up of the individual. Our genotype is present from conception and never changes.

Phenotype is the observable characteristics that are produced by that genetic endowment. Phenotypes can be affected by other genes and by the environment (such as climate, diet and lifestyle).

Many kinds of variation are influenced by both genetic and environmental factors. Though our genes govern what characteristics we inherit, our environment can affect how these inherited characteristics develop. For example, an individual may have inherited a tendency to be tall but a poor diet during childhood will cause poor growth.

Genetic Conditions

We've seen that each human has 23 pairs of chromosomes and at conception we inherit one chromosome from our mother and one from our father making an individual pair. Genes are segments in the chromosome and are your body's instruction manual. Some genes are dominant and some recessive. Let's look at the example of hair colour to explain the difference.

Brown hair is regulated by a dominant gene, blond hair by a recessive one. My father has brown hair and my mother has blonde hair; I have brown hair, yet my brother has blond hair. How is this so? If at conception the gene for brown hair is present the child will have brown hair because it is the dominant gene and 'overpowers' the recessive blond hair gene. However, even though my father has brown hair he could also carry the blond hair gene (his mother had blonde hair). Thus at the moment of my brother's conception my mother and father both passed down the recessive blond gene and this is why my brother has blond hair.

Genetic inheritance is rather like the Lotto: you never know the mix of numbers you might end up with. Conception is literally a lottery!

Certain conditions and diseases are genetically inherited.

Figure 3.1: Inheritance pattern of sickle-cell disease

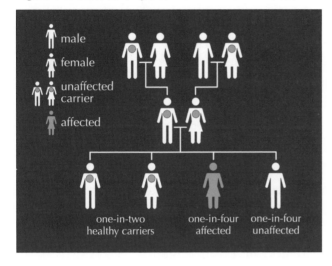

Recessive Inheritance

If two parents are both carriers of a genetic condition with a recessive inheritance pattern, there is a one in four chance that each child will be affected. So on average, one-quarter of their children will be affected. There is also a one in two chance that each child will be an unaffected carrier, like the parents. Examples of genetic conditions that show a recessive pattern of inheritance are cystic fibrosis, sickle-cell disease, Tay-Sachs disease and phenylketoneuria (PKU).

Figure 3.2: Inheritance pattern of Huntington's Disease

Dominant Inheritance

If one of two parents is affected by a genetic condition with a dominant inheritance pattern, every child has a one in two chance of being affected. So, on average, half their children will be affected and half their children will not be affected and so will not pass on the condition. However, as chance determines inheritance, it is also

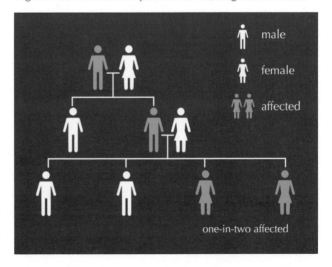

possible that all or none of their children will be affected. Examples of genetic conditions that show a dominant pattern of inheritance include Huntington's Disease.

The major difference between dominant and recessive gene inheritance is as follows: **Recessive**. For the child to inherit the condition it must inherit both variant (affected) genes from both parents. Thus a child has a one in four chance of inheriting the condition, a one in two chance of becoming a healthy carrier and a one in four chance of not inheriting the affected gene.

Dominant. One parent is affected and the other parent is healthy. At conception, if the variant gene of the affected parent is transferred the child will inherit the condition. It does not matter if the other parent's genes are healthy. Thus with dominant conditions such as Huntington's Disease, the child has a one in two chance of inheriting the condition.

Chromosomal Disorders

We have looked at gene-linked disorders including cystic fibrosis, PKU and Huntington's Disease. These are different from chromosomal abnormalities such as Down Syndrome. Both occur at conception, but whereas the gene-linked conditions are inherited, chromosomal conditions generally occur through the presence of an extra chromosome. Each cell has 23 pairs of chromosomes. In Down Syndrome, an extra copy of chromosome 21 is present. Sometimes you will find Down Syndrome referred to as 'Trisomy 21', the 'tri-' referring to the three chromosomes that are present instead of two. Other chromosomal syndromes include Patau Syndrome (Trisomy 13) and Edwards Syndrome (Trisomy 18). Sex-linked chromosome abnormalities can also occur (Fragile X Syndrome).

Down Syndrome

Down Syndrome is the most common of the chromosomal disorders. In Ireland, it is estimated that one baby in 600 is born with Down Syndrome. Physical characteristics associated with Down Syndrome include a downward sloping skin fold at the inner eye corners, flat appearance of the face, protruding tongue and low muscle tone (which means that a baby's head and neck will need extra support). Down Syndrome is accompanied by developmental delay; IQ can range from low to severe learning disability.

'Share the Journey' is the motto of Down Syndrome Ireland. Their mission statement says:

> People experience many great things and also face many challenges throughout their lives. People with Down syndrome are no different, but may need a little

extra help and support along the way. Down Syndrome Ireland's goal is to help people with Down syndrome make their own futures as bright and independent as possible by providing them with education, support and friendship every step of the way.

It is really important that people with disabilities are seen primarily as individuals — who happen to have a disability.

Compare and Contrast

Down Syndrome Ireland states that 'the quality of life of people with Down syndrome has improved immensely in the last thirty years. Just like the rest of us, they now enjoy longer life expectancy and can live happy, fulfilling and active lives as part of the community.'

What do you think was the experience of a child born with Down Syndrome in the 1940s? Do you think their experience would be different from that of a child with Down Syndrome born in Ireland today? (See the end of Chapter 2 for some ideas.)

PRENATAL DEVELOPMENT

Prenatal development consists of three stages:

- the germinal stage
- the embryonic stage
- the foetal stage.

The **germinal stage** is approximately the first two weeks of development. When a woman's egg (ovum) is fertilised by a man's sperm, the fertilised egg is called a zygote. Once fertilised, a process of continuous cell division begins. Approximately 10 to 14 days after conception the zygote, which now contains a mass of cells, attaches itself to the mother's uterus.

The **embryonic stage** is from the end of the second week to the eighth week. The cell mass is now called an embryo. The placenta and umbilical cord begin to develop during this stage and the bodily organs and systems begin to form. By week eight the heart is beating and the brain is forming. Facial features such as the eyes become discernible.

The **foetal stage** runs from the ninth week until birth. The embryo is now referred to as a foetus. At 24 weeks the eyes open. At 28 weeks the foetus attains the age of viability, which means that it is likely to survive outside the womb in the case of premature birth (Hetherington & Parke 1999).

Figure 3.3: Critical periods in human development

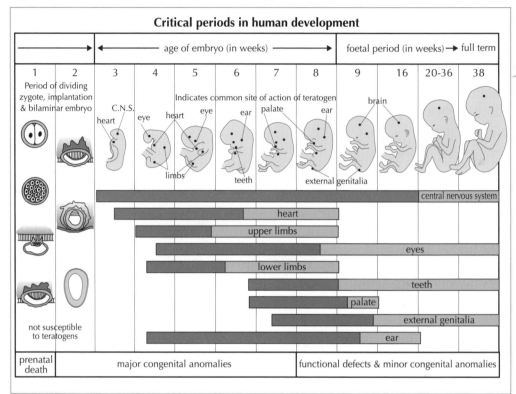

Hazards During Prenatal Developments

As we saw earlier, chromosomal and genetic disorders affect the foetus at the moment of conception.

Other external factors can affect and influence the foetus's development in the womb. These influences include diet and nutrition: if a pregnant mother is not getting enough calories, the foetus is more likely to have a low birth weight compared with a pregnant woman who gains 26 pounds or more. Factors such as age and emotional stress have been implicated in prenatal development as well as certain illnesses.

Approximately a quarter of infants born to HIV-infected mothers become infected. The virus crosses the placenta and enters the foetus's bloodstream or it can be transferred in the birth canal during birth. Research has found that HIV-positive mothers who take the drug AZT have a lower chance of passing the disease to their children.

Rubella or 'German measles' is most dangerous if contracted in the first weeks of pregnancy. It can cause hearing impairments, cataracts and heart defects.

Teratogens

Some external influences, called teratogens, can cause birth defects. The timing of exposure and dose of a teratogen can make a difference in outcome. Take the example of thalidomide. This was a drug available from chemists that was used to ease morning sickness and was recommended to pregnant women in the late 1950s. It became apparent as the pregnant women who had taken the drug began to give birth that their babies had been affected. The infants were born with missing arms and legs or shortened limbs. Figure 3.3 shows that there are periods when different parts of the body are more vulnerable to teratogenic exposure. These women were taking the medication for morning sickness, which can occur early in pregnancy, just at the stage when limbs are vulnerable.

As we examine different teratogens bear in mind that timing, dose and duration of exposure all interact together in determining outcomes. Another factor can be genetic vulnerability, which we'll explore more in the focus piece on Foetal Alcohol Syndrome Disorder in Chapter 4. Some other teratogens are cigarettes, alcohol (see Chapter 4) and certain illegal drugs. Smoking during pregnancy increases the risk of miscarriage, premature birth and low birth weight.

According to Boyd and Bee:

> The infants of twice-weekly marijuana smokers suffer from temors and sleep problems. They seem to have little interest in their surroundings for up to 2 weeks after birth ... both heroin and methadone ... can cause miscarriage, premature labour, and early death. (2005:67)

They continue that sixty to eighty per cent of babies born to heroin- or methadone-addicted mothers are addicted themselves and suffer withdrawal symptoms such as high-pitched cries and convulsions.

BIRTH

The beginning of labour is caused by a hormonal response that encourages contractions to start. Contractions cause the cervix to dilate (open), allowing the baby to move down the birth canal (vagina) and be born. The mucus plug is expelled and the waters break (sometimes this is done deliberately at the hospital if labour is progressing slowly). The process of labour consists of three stages.

Stage 1 consists of contractions, which can be 15 or so minutes apart. The purpose of these contractions is to open the cervix, which is at the neck of the womb. A typical pattern is to experience one contraction of about 40 to 50 seconds every ten minutes at the beginning of stage one, and one contraction lasting more than a minute and coming every 30 seconds or so at the end. By the end of the first stage of labour the cervix will be fully dilated at ten centimetres.

Stage 2 begins when the baby's head starts to move down the birth canal. This phase is marked by the mother actively pushing the baby down the birth canal. This process can last anywhere between several minutes and two hours.

Stage 3 involves the delivery of the afterbirth including the placenta, umbilical cord and other membranes. This is the shortest stage of labour.

MEASUREMENT AND ASSESSMENT

Apgar Scale

The Apgar Scale is used in the first minutes of a baby's life to assess its health. It is highly effective because it's quick and simple to do. It is administered within the first minute after delivery and then repeated five minutes later. The newborn is assessed on the following:

1. Heart rate: the strength and regularity of the heart beat. A hundred beats a minute scores 2, less than 100 scores 1 and no heartbeat scores 0.
2. Respiration: regular breathing or crying scores 2, irregular breathing scores 1 and no breathing scores 0.
3. Muscle tone: a score of 2 is given for active movements, some movement is scored at 1 and no movement is scored 0.
4. Reflex response: strong reflexes score 2, weak reflexes 1, and no reflex 0.
5. Skin colour: this shows how oxygenated the blood is. A score of 2 is given for pink, blue extremities are scored at 1 and totally blue skin scores zero.

Each of these factors is given a score between zero and two and then added up. Most infants score between 7 and 10, with 10 being a perfect score.

- 8–10: excellent condition
- 5–7: fair condition; the baby might need some help with breathing
- under 5: baby requires treatment; a paediatrician may be called to assess him/her.

It should be borne in mind that the Apgar Scale is only intended to measure the health of the baby on delivery and is not a predictor of long-term health.

PKU

In the section on conception we encountered a genetically inherited condition called phenylketonuria (PKU). This is a condition that if diagnosed can be treated and controlled, but if undetected it can lead to severe mental retardation because amino acids build up and cause brain damage in the child. It can be treated with a specially modified diet that excludes the specific amino acid, thus preventing it from accumulating in the child's system. Within the first few days of life a pin prick is made on the infant's heel and blood extracted for analysis and diagnosis. According to the *National PKU News* (Fall 2001), since testing began in Ireland there has been an incidence rate of 1 in 4,500, which is far greater than the incidence rate of 1 in 12,000 seen in the USA. They reported that in 2001 611 individuals with PKU (334 males and 277 females) attended the National Centre for Metabolic Disorders at the Children's Hospital in Temple Street, Dublin.

Table 3.1: Incidence of congenital disorders in Ireland and worldwide

Condition	Date Started	Irish Incidence	Worldwide Incidence
Phenylketonuria (PKU)	1966	1:4,500	1:12,000
Homocystinuria (HCU)	1971	1:65,000	1:200,000
Classical Galatosaemia	1972	1:19,000	1:60,000
Maple Syrup Urine Disease (MSUD)	1972	1:125,000	1:216,000
Congenital Hypothyroidism	1979	1:3,500	1:4,500
Congenital Toxoplasmosis (2-year pilot programme)	2005		

Newborn Screening in Ireland

The National Newborn Screening Programme was first established in the Republic of Ireland in 1966 to screen all infants for PKU. It was one of the first national programmes in the world. Since then it has expanded its services and it now tests for six disorders, including congenital toxoplasmosis (under a pilot programme that began in 2005). The disorders screened for at present are:

- phenylketonuria (PKU)
- classical galatosaemia
- maple syrup urine disease

- homocystinuria
- congenital hypothyroidism
- congenital toxoplasmosis.

For more information on the National Newborn Screening Programme see www.nnsp.ie/pdf/disorders.pdf.

Low Birth Weight Babies

- Low birth weight: under 5½ pounds (2,500 grams) at birth.
- Very low weight: under three pounds (1,360 grams) at birth.
- Extremely low weight: under two pounds (907 grams) at birth.

Preterm babies are born before 38 weeks' gestation. Most preterm babies are low-weight.

Small for date infants are babies whose birth weight is below normal when the length of pregnancy is considered and who weigh less than 90% of all babies of the same gestational age.

Cigarette smoking is the leading cause of low birth weight in infants (Unicef 2004). Adolescents who give birth when their bodies have not fully developed are at risk of having low birth weight (LBW) babies.

Irish Birth Statistics

Some figures from the Irish National Perinatal Reporting System in 2002:

- average birth weight: 3,474 grams (approx. 7.66 pounds)
- average gestational (in the womb) age at delivery: 39.5 weeks
- low birth weight babies (weighing less than 2,500 g or 5.5 pounds): 4.9% of all births.
- between 1993 and 2002 LBW babies as a proportion of all live births increased by 18.2%
- there were 864 twin births, 18 triplet and 3 quadruplet births
- average age of mothers giving birth: 30
- single mothers accounted for 30% of all women giving birth (an increase of 65.4% since 1993)
- delivery by Caesarean section: 22.4% of all live births.

In Focus: Low Birth Weight and Social Inequality

A report entitled *Unequal at Birth* was published by the Institute for Public Health in Ireland (McAvoy *et al.* 2006). The authors show that social inequalities relate to the incidence of low birth weight births. In the report the indicator used for low birth weight babies was a weight less than 2,500 grams (5½ pounds).

Some of their findings were as follows:

In 2000, 55,166 births were notified to the National Perinatal Reporting System (NPRS) in the Republic of Ireland. The average birthweight of babies born in that year was 3,491 grams. There were an estimated 2703 low birthweight babies recorded in 2000, representing 4.9% of all births. Between 1991 and 2000 the proportion of low birthweight babies has increased by 15.1% ... It is not possible to conclude whether this increase observed in the proportion of low birthweight babies born is significant, representing a true increase, or whether it could be attributed to improvements in the NPRS data collection system since 1998. (p.20)

Certain women are especially vulnerable to poverty and social exclusion and are therefore likely to suffer health inequalities in relation to their pregnancy and their babies. These groups were identified and include teenage mothers, lone parents, disabled women, Travellers, refugees and asylum seekers, other ethnic minority women, women prisoners and homeless women. (p.28)

From observational studies conducted in Ireland and from previous annual reports of Ireland's National Perinatal Reporting System, it is evident that women in lower socio-economic groupings are more likely to experience a teenage pregnancy and, as young mothers (aged 19 years or less), they are more likely to deliver a low birthweight baby. Teenage pregnancies are associated with prematurity and the preterm delivery rate of teenagers far exceeded matched controls of women aged 20–24 years in one Irish maternity hospital.

Almost ninety per cent of attendees at a Dublin Adolescent Antenatal Booking Clinic were recorded as being in the lowest socio-economic group and most had poor educational attainment. Eighty-seven per cent of these mothers had left school. Eighty per cent of them had not sat the Leaving Certificate with another ten per cent reporting never sitting a state examination. A quarter of these women were over twenty weeks pregnant at first presentation to health services with over two thirds saying they were afraid to attend hospital earlier ... It is also well recognised that pregnant schoolgirls are more likely to live in areas with poor housing, overcrowding and high unemployment rates. (p.23)

CHAPTER FOUR

THE NEWBORN BABY

PHYSICAL ASPECTS

In the first few days, an infant will lose between five and seven per cent of their total body weight. This is normal and reflects their new adjustment to feeding outside the womb.

Reflexes

Infants are born with a number of instinctual responses to stimuli (for example touch, bright light). These responses are called **primitive reflexes**. These reflexes develop while the infant is in the womb and are present at birth. We are born with these reflexes — a newborn has no control over voluntary movements of the body. They gradually diminish as higher functioning parts of the brain develop, enabling the brain to take control as motor (movement) development increases. However, we do retain some of these reflexes, including the gag and blink reflexes. The reflexes discussed below are those that disappear in the first year of life.

Swallowing and Sucking

When anything is placed in the mouth the baby will automatically suck and swallow. These reflexes make sense — they ensure that the baby is primed and ready for feeding. They reflexes are common to all mammals and aid feeding. The sucking and swallowing reflexes disappear after approximately four months.

Rooting

Touching an infant on the side of the cheek will cause them to turn to that side in search of the breast. If you touch a baby's cheek the baby will turn in the direction of the touch and the mouth purses searching for the nipple. As with sucking and swallowing this behaviour can be seen as assisting feeding.

Walking or Stepping

If a baby is held upright with their feet on a firm surface and slightly tilted forward, they will attempt to make stepping movements.

Palmar Grasp

The infant tightly closes its fingers when pressure is applied to the palm of the hand by a finger or an object. When an object is placed in the palm of the baby's hand it will grasp it. This grasp is incredibly strong. This reflex continues for approximately five or six months after birth.

4.1: The stepping reflex

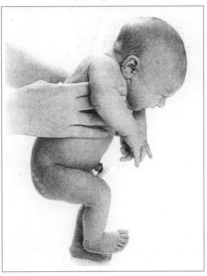

Moro or Startle Reflex

This reflex is observed when a baby thinks it's about to fall. They will throw their arms out, their palms open, and make a grabbing movement by bringing their arms back into their chest, almost as if they're trying to catch hold of something. (If you were falling, you would probably throw your arms out and try to grasp something to stop yourself from falling.) This reflex generally disappears between three and four months of age.

Reflexes are an important indicator of the health of the baby because they are governed by the central nervous system. If a reflex is absent or impaired it indicates that further investigation is warranted.

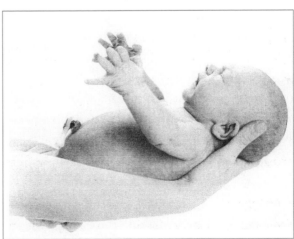

4.2: The Moro or startle reflex

SENSATION AND PERCEPTION

- **Smell** – neonates can differentiate odours and recognise mother's odour at two days old (MacFarlane 1975).
- **Hearing** – a baby can distinguish their mother's voice from that of another female (DeCasper & Spence 1986).
- **Sight** – Neonates can distinguish their mother from another female at five days old (Bushnell *et al.* 1989).

MATERNAL BONDING THEORY

Herbert (2002:54) comments that during the 1970s a stark 'critical period' for bonding was proposed. The idea was that during the first hours after birth the mother engaged in bonding with her baby through tactile, visual and olfactory stimulation. It was suggested that if this bonding was interrupted (if, for example, the baby was removed to intensive care) it could have negative long-term effects on the mother–child relationship. Thankfully this stark belief isn't popular any more but maternal bonding still holds interest in many quarters. Studies have found no evidence that disruption of 'close contact' between mother and child has any impact on their relationship or mothering effectiveness.

WHAT NEWBORNS CAN DO

Newborn babies:
- can focus on objects less than one metre away
- show a marked preference for human faces
- can recognise their mother's voice
- use their senses to explore their new environment (seeing, hearing, smelling, tasting and touching)
- make eye contact and cry to indicate need
- are often able to imitate adults, e.g. copying adults sticking out their tongue.

According to Meggitt (2006:11), the following activities will promote development:
- provide lots of physical contact and maintain eye contact
- talk lovingly to babies and give them the opportunity to respond
- pick babies up and talk face to face with them
- use bright, contrasting colours in furnishing
- feed babies on demand, expect no set routine within the first few weeks.
- encourage babies to lie on the floor and kick and experiment safely with movement.

In Focus: Foetal Alcohol Syndrome — the Irish Experience
According to Rich (2005:12) foetal alcohol syndrome disorder is an umbrella term describing the 'continuum of complex neuropsychiatric, cognitive, behavioural and physical symptoms associated with all levels of prenatal alcohol exposure'. Fetal Alcohol Spectrum Disorders Ireland (www.FASD.ie) is a group set up by people who have come in contact with children with foetal alcohol spectrum disorders. Their website outlines the following characteristics of foetal alcohol damage.

A child with foetal alcohol syndrome

How does prenatal exposure to alcohol affect a child?
Alcohol crosses the placenta undiluted, and within minutes the level of alcohol in the foetus's blood reaches maternal blood alcohol level. Alcohol is a teratogen which can cause:
1. physical malformation (alcohol exposure has been linked to cases of spina bifida, heart defects, kidney defects, etc.)
2. learning and behavioural challenges.

Babies with foetal alcohol syndrome (FAS) (with confirmed maternal alcohol exposure) showed evidence of:
3. a characteristic pattern of facial anomalies with features such as short eye openings, a thin upper lip, low nasal bridge, flattened philtrum and a flat midface
4. low birth weight for gestational age, deceleration of weight gain over time which is not due to nutrition issues, and disproportional low weight to height
5. central nervous system neurodevelopmental abnormalities, as in at least one of the following;
    • decreased cranial size at birth
    • structural brain abnormalities such as microcephaly, partial or complete agenesis of the corpus callosum, cerebrellar hypoplasia
    • neurological hard or soft signs (as age appropriate), such as impaired fine motor skills, neurosensory hearing loss, poor tandem gait, poor hand-eye co-ordination. (Synopsised from Stratton *et al.* 1996:76–7.)

Other risks are:
1 partial foetal alcohol syndrome (pFAS)
2 alcohol-related neurodevelopmental disorder
3 alcohol-related birth defects
    (Ryan & Ní Chionnaith (2005).

How Common is FASD?

There are no official statistics available for the prevalence of FASD in Ireland. Applying the rates of FAS reported by the Substance Abuse and Mental Health Services Agency, Center for Disease Control in the United States (SAMHSA 2005) to the CSO's figures for Irish birth rates in 2002, we can estimate that one per cent (605 out of the 60,503 babies born in Ireland in 2002) could have FASD. The incidence may be higher due to the high rate of binge drinking among young women in Ireland.

No two children with FASD are exactly alike, either behaviourally or physically. Some of the co-occurring behavioural, social and learning characteristics may include:

1. Attention problems or hyperactivity (Morse 1991; Nanson and Hiscock 1990).*
2. Academic problems, including specific deficits in mathematics and memory skills (Streissguth *et al.* 1993).*
3. Very specific language deficits (Abkarian 1992).*
4. Problems with adaptive functioning that grow more significant with age (Lemoine & Lemoine 1992; Streissguth & Randels 1989).*
5. Behavioural challenges.
6. Social or relationship challenges including difficulty making or sustaining friendships.
7. Sensory impairments such as vision or hearing.
8. Sensory integration challenges including auditory, visual and tactile processing.
   *SAMHSA 2005, cited in Ryan & Ní Chionnaith 2005:4.

In the USA a lot of attention has been paid to the effects of alcohol on the growing foetus. In Ireland medical professionals are more sanguine: while their advice would be to abstain from alcohol during pregnancy, low amounts of alcohol are not considered problematic. (See a discussion on this subject with Dr Peter Boylan, consultant obstetrician, and Tom Donaldson, President of NOFAS (National Organisation Fetal Alcohol Syndrome) at www.fasd.ie/books_frame_page.html.)

This section is only an introduction to the topic of foetal alcohol syndrome. It is not a comprehensive guide nor does it represent all issues in this debate. It is a starting point and a source of information for those interested in this issue.

CHAPTER FIVE

INFANCY: THE FIRST TWO YEARS OF LIFE

The first two years of life is a time of immense development and growth. This is very obvious in the realm of physical development: within a year an infant will generally be able to crawl, stand and begin to walk. Within a lifespan perspective it can be argued that these years are particularly significant for future functioning in later life. Within a nature–nurture context, we will look at the role and impact of 'nurture' (environment) on a child's development.

CHAPTER OUTLINE

- Physical development:
  - Patterns of growth
  - The brain
  - The central nervous system
- Perception and sensation:
  - Sensations
  - Perception
- Cognitive development:
  - Piaget's sensori-motor stage (0–2 years)
- Socio-emotional development:
- Freud: the oral stage (0–2 years)
- Erikson
- Attachment:
  - Development of attachment theory
  - John Bowlby
  - Mary Ainsworth
- The development of 'self':
  - Implicit self-knowledge
  - Personality and temperament
- In Focus: Community Mothers Programme
- Chapter summary

PHYSICAL DEVELOPMENT

Different patterns of growth in the body occur at different stages of development. In the first two years we witness an **asynchronous** (very rapid and uneven) period of growth. Development includes changes in weight, height, muscle growth, length of bones, laying down of fat, and the growth of internal organs such as the heart, lungs, brain and nervous system. In the first year of life, a baby will generally grow 25–30 cm in length and treble their body weight.

Patterns of Growth

Cephalocaudal

The **cephalocaudal** pattern of growth refers to growth developing from the head down. At birth a baby's head is disproportionately bigger than the remainder of the body. This is because the baby's head grows rapidly just before birth. After birth the rest of the body catches up. This can be seen in the motor development of the child. Within the first few months a baby if laid on its stomach in the prone position will attempt to raise itself up: this reflects the growth of and increased strength in shoulders and arms. As growth continues down the body in accordance with the cephalocaudal principle, the baby will begin to crawl as strength and growth in the lower half of the body now develop.

Proximodistal

The **proximodistal** pattern of growth refers to growth from the trunk to the extremities. This is most easily recognised in infants as they acquire skills involved in grasping and manipulating objects. Think of a young infant attempting to hold a cup: initially the infant will grasp the cup with both hands, locking the palm of their hands around the object. As they develop they will gain more control of their fingers until they are finally able to pick up the cup with their fingers. It's similar to learning to hold a pencil. Young children will grasp the pencil in the palm of their hand but as they develop, in line with the proximodistal principle, the continued development in their fingers will enable them to hold the pencil between their fingers and thumb with greater control.

The Brain

The brain at birth is 25 per cent of its eventual adult weight. By the child's second birthday the brain will have increased to 75 per cent of its adult weight. This gives us some insight into the rapid changes that occur in the brain during the first two years of life. In the previous chapter we looked at the reflexes that an infant is born with and how these diminish as the brain develops and takes over from these reflexes. The brain is involved in all aspects of development: motor (movement), sensory, cognitive and social. Before we look at the changes experienced in the brain we need to learn a little more about structures of the brain.

Figure 5.1: The four lobes of the brain

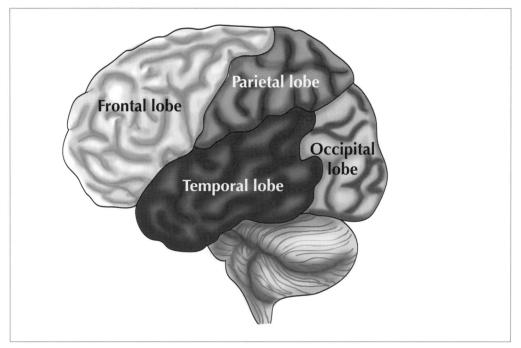

The brain consists of four lobes. These lobes have different locations and functions:

- **Frontal lobe.** Located behind the forehead, it is the 'front' of the brain. This lobe does not fully develop until the teen years. Its functions include higher forms of thinking, such as reasoning, and voluntary movements.
- **Parietal lobe**. Located behind the frontal lobe, around the top of the head. Functions include processing information about body sensations.
- **Occipital lobe.** Positioned at the back of the brain. One of its main functions is vision.
- **Temporal lobe.** Located at the side of the brain. Involved in hearing.

The brain is not one huge mass; it is divided into two halves or **hemispheres.** These are connected by a bundle of nerves called the **corpus callosum** that allows each half to communicate with the other, thus ensuring that they are working in unison.

Lateralisation is the term that's used to describe the specialisation of function in one or the other hemisphere. It means that one hemisphere can have a particular function for which it is responsible. For example, in most people the left side of the brain is generally responsible for language. However, this does not mean that the right side has no role at all in language: it does have a part to play, but not the dominant role. It is generally accepted that lateralisation occurs between the ages of two and twelve.

The Infant Brain

An infant is born with all the brain cells it will need — in fact it is born with more than it needs. As the brain develops it will cut back or 'prune' cells and pathways it does not need or use.

Imagine a city that has hundreds of little roads. It comes to the mayor's attention that some of these roads are not very efficient or are not being used and that an overhaul is needed. So he orders that the inefficient or superfluous roads that aren't contributing to the smooth running of traffic are removed. When this is done the traffic moves more smoothly and quickly as it is easier and clearer now to get from one point to another. So although there are fewer roads, the roads that are left are the ones that work best and ensure traffic flows quickly and efficiently.

This is similar to how the brain operates in pruning away superfluous neural pathways and cells; it is ensuring the brain works more efficiently.

Interestingly, the interaction of the child with its environment influences the development of the brain. We saw in the 'In Focus' section in Chapter 1 that the brain needs a stimulating environment. If a child is severely neglected the brain does not develop normally. Pathways for language can be lost if no stimulation or access to language is available to the child. In fact a neglected child's brain can weigh significantly less than the brain of an average child of similar age. So the child's environment can influence the development of the brain.

Another process that occurs during the first two years of life is **myelinisation**: the nerve fibres are insulated, thereby improving and speeding up nerve impulses.

The Central Nervous System

The central nervous system is made up of the brain and spinal cord. Its main function is to relay information to and from the brain. A healthy nervous system improves co-ordination of the limbs and movement.

Table 5.1: Motor and language development 0–2 years

Age	Gross Motor Skills	Fine Motor Skills	Language
3 months	• Can lift head and chest into prone position • Can wave arms and kick legs	• Plays with their fingers • Can hold a rattle for a short time	• Coos and laughs • Responds to sounds • Plays with speech sounds
6 months	• Can roll onto stomach if placed on back • If held standing or sitting can do so with a straight back	• Palmar grasp: will use whole hand to pass object from one hand to another • Will put objects in mouth to explore them	• Will make babbling sounds • Attempts to match what they hear
9 months	• Can sit without support for 15 minutes • Stands if holding onto furniture • Starts crawling	• Pincer grasp: grasps objects between finger and thumb • Can manipulate objects	• Begins to understand words, e.g. 'no' and baby's name
12 months	• Can stand unsupported for a short while • Can walk with one hand held	• Can hold crayon in a palmar grasp • Shows hand preference • Builds with bricks	• Around this time, says first words • Understands function of naming things
18 months	• Can climb into an adult chair • Climbs up and down stairs with support • Can go down stairs on their stomachs alone	• Can build tower with 3 or more blocks • Can use a spoon to feed themselves • Can use the pincer grasp to pick up very small objects	• Says first sentence consisting of 2 words • Refers to themselves by name • Echoes the last part of what others say (echolalia)
24 months	• Can run safely • Walks up and down stairs • Can push and pull wheeled toys	• Can build tower of 6 or more blocks • Drinks from a cup • Draws circles and lines	• Expanding vocabulary • No longer babbles, wants to talk

Adapted from Meggitt (2006)

Shaken Baby Syndrome

We've looked at how rapidly the brain develops. Young infants have weak neck muscles and are unable to support the weight of their heads. Furthermore, the blood vessels in the developing brain are fragile. For these reasons it is essential that the baby's head is protected and care is taken to prevent falls, etc. A form of abuse that can involve damage to the infant's brain is **shaken baby syndrome**, which involves the violent shaking of the infant resulting in damage to the brain. Injuries can include blindness, seizures, developmental delay, brain damage and death.

PERCEPTION AND SENSATION

According to William James (1890), 'the world of the newborn is a buzzing, blooming confusion, where the infant is seized by eyes, ears, nose and entrails all at once'.

Over a hundred years on and a lot of research later we now know that babies are more competent than first imagined and better equipped to make sense of their new world. It is not quite the 'buzzing confusion' first thought of by James.

Sensation refers to the functions of the sense organs (vision, hearing, taste, touch and smell), the awareness of conditions or feelings inside and outside the body.

Perception means interpreting the sensation experienced and giving it meaning.

Sensations

Vision

This is one of the least developed systems in the newborn baby. The newborn's abilities develop quickly, and within a few days they can see objects about a foot away. Infants also display visual preferences (see the discussion of Fantz's experiments on p. 54).

Hearing

According to Aslin (1987) babies show a preference for high-pitched voices more typical of a female voice and will turn their head in the direction of sounds made. DeCasper and Spence (1986) found that infants display a preference for familiar sounds that were transmitted through the womb in the final months of pregnancy. This finding further contributed to the growing evidence of prenatal learning.

Taste

Babies show a preference for sweet substances and will grimace if they taste bitter or sour flavours. Explanations offered for this preference refer to the fact that breast milk is sweet, suggesting that the infants are primed to take the milk. Toxic or poisonous substances tend to have a bitter taste so it is arguable that infants have an innate aversion to these dangerous substances.

Smell

Week-old infants have been shown to indicate an odour preference for the mother's breast milk when presented with pads of several other mothers' milk (MacFarlane 1975).

Touch

Newborns respond to touch and to changes in body temperature.

Perception

Gibson's Depth Perception Experiment

Eleanor Gibson, an eminent psychologist, has spent her long career investigating perceptual development and is well known for her investigation of depth perception using the visual cliff experiment (Gibson & Walk 1960). As you can see from the image Gibson simulates a cliff or drop using a table with a glass top to give the impression that it is a cliff. Most infants will refuse to crawl beyond the apparent cliff edge, despite the

Figure 5.2: Gibson's cliff experiment

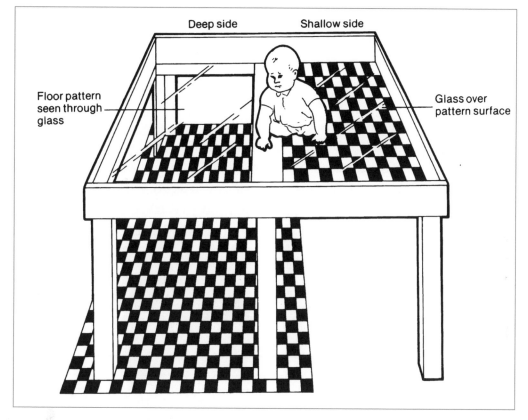

coaxing of their mothers. This has also been seen with young animals. Gibson took an ecological approach to her theory of perception and saw children as active participants in their environment. She further believed that the human species has evolved adaptive ways of perceiving their environment.

Gibson incorporated the idea of 'affordances', a term that is used to describe the opportunities an environment holds for learning and adaptation. An example of this would be a hill that a baby has to try to climb and negotiate, thus the hill (or environment) offers an opportunity for the baby to learn something new. Gibson was interested in *how* infants, in such an information-laden world, decide what pieces of information to ignore and which to take on board. Indeed children have been described as the 'hunters and gatherers' of the information world. Gibson argues that when children interact with their environment it enables them to adapt to it, for example when they perceive faces or play with flowers this provides an opportunity for adaptation. Gibson focused on the question 'How do we learn to perceive our environment?' and her response was that children become increasingly selective as they develop and learn to detect information that identifies events and objects that they can use for their daily activities.

Face Perception – Fantz's Experiment

Fantz (1961) devised an experiment to test whether infants showed a preference for human faces. As you can see from the image he showed them pairs of faces, a discernible face, a scrambled up face and a blank face. These were shown to infants ranging in age from four days old to five months. Fantz used a technique called 'preferential looking', which involved measuring the length of gaze of the infant to determine which image was of most interest or was most familiar to them. He found a definite preference for the 'realistic' face. Fantz's research suggests that the ability to recognize faces is either innate or learned very quickly after birth. Thus not only does Fantz's research shed light on infant's face perception but it also adds to the debate about whether perception is innate or learned.

Figure 5.3: Faces used in Fantz's study

Nature or Nurture?

Considering Fantz's findings along with Gibson's depth perception, do infants have certain innate abilities, such as not crawling off the cliff or being able to perceive a human face, or do they learn these abilities after birth? What do you think?

COGNITIVE DEVELOPMENT

In Chapters 1 and 2 we discussed what is meant by cognitive development and theories related to it. Now we'll look at theories of cognitive development at specific age periods or stages.

Piaget's Sensori-motor Stage (0–2 Years)

Features of this stage include:

- object permanence
- causality
- representational ability
- imitation.

We saw in Chapter 2 that the sensori-motor stage represents the first stage of Piaget's cognitive theory. The title gives us a clue as to how learning is accumulated during this period. During this stage we witness the use of the senses and movement by the infant to learn about their world and to construct knowledge of it.

As they develop, infants realise that they can make things happen — they bang spoons and throw objects from their high chair. This also reveals that the infant sees themselves as separate from their environment and that they are developing a sense of agency about their actions, in other words, that they are the 'agent' or initiator of their actions. Another aspect to this is **causality**, when the child begins to recognise causal relationships: for example, if I shake this rattle it will make a noise.

Object Permanence

Definition: object permanence means the understanding that objects continue to exist even when we can no longer see them.

Its acquisition towards the end of this stage is considered an important milestone in cognitive development.

A simple test for object permanence is to show a child a toy and then throw a blanket over it. If the child seeks the toy out it is an indication that they understand that the toy continues to exist even when they can no longer see it. Related to object permanence is the concept of **representational ability**. This involves the ability to make a 'mental representation' of an object, i.e. to be able to represent it mentally or in your mind. This allows the child to manipulate the object, which can lead to the further acquisition of problem-solving skills.

A good way to understand representational ability or mental representation is to imagine you're driving to college/work and you get stuck in a traffic jam. If you know the area you can mentally call up a map of it and start figuring out a new way to get to your destination. You map out your new route in your mind before you commit to it. The ability to map the area in your mind reflects representational ability. Piaget believed it develops between 18 and 24 months of age.

Imitation

Imitation is the reproduction of an expression or of a behaviour. Piaget proposed that imitation was achieved at roughly nine months of age and that deferred imitation occurred later, after the acquisition of mental representation.

Meltzoff and Moore (1977) found that within a few weeks of birth babies could imitate or reproduce a facial expression made by an adult, such as sticking their tongue out. Some have argued that this is merely a reflex, but Meltzoff and Moore believe this reflects a biologically based capacity to imitate.

Meltzoff (1988) also found evidence of what he believes to be a nine-month-old's capacity for deferred imitation. This means that the infant reproduces a gesture after a certain amount of time has elapsed, for example the following day. This is noteworthy when we consider that Piaget did not believe that children acquired 'object permanence' until the end of their second year. In order to have object permanence the child must be able to make a 'mental representation' of the object, that is, they must be able to represent it mentally so that they know the object continues to exist even though they can no longer see it. Yet in order to imitate a facial expression these very young infants must already possess these abilities to some degree. Not only does Meltzoff and Moore's research demonstrate that 'imitation' is acquired earlier than Piaget proposed, their research would suggest that infants acquire concepts such as object permanence and mental representation earlier than proposed by Piaget.

Check! Learning Goals

You should be able to:

- identify Piaget's first stage of cognitive development
- describe briefly 'object permanence'
- explain 'imitation'
- outline Meltzoff and Moore's research and discuss its implications in relation to Piaget's theory.

SOCIO-EMOTIONAL DEVELOPMENT

Freud: The Oral Stage (0–2 Years)

The mouth is the prime source of pleasure, and Freud believed that feeding and more particularly the weaning period should be handled with care. He advised against weaning a baby too early or indulging the infant and weaning too late — doing either could result in the infant becoming 'fixated' at the oral stage. This would have negative consequences in later life, including swearing, smoking, biting fingernails, overeating, etc. All these behaviours are related to the mouth area. Freud advised that through oral satisfaction, the baby develops trust and an optimistic personality. If the child experiences oral dissatisfaction the personality may become pessimistic, aggressive and distrustful.

Check! Learning Goals

You should be able to:

- name Freud's first stage and the age group to which it applies
- explain what is meant by 'oral fixation'.

Erikson

Trust vs. Mistrust (0–1 Year)

In the first year of life the infant is completely dependent on its caregivers. How well the infant's needs are met and how sensitive the parenting is determines whether the infant develops trust or mistrust of the world.

Practical applications:

- holding babies close and having warm physical contact with them when they are being fed

- responding to them immediately when they are distressed or crying: this can promote a sense of trust (Garhart Mooney 2000:41).

Autonomy vs. Shame and Doubt (1–2 Years)

Children begin to walk and assert their independence. As the term 'autonomy' suggests, children can come to believe in themselves and their abilities through the encouragement and support of their caregivers. If the child's efforts are ridiculed, the child will develop a feeling of shame and doubt. Toilet training is seen as one of the key events which can influence the outcome of this stage.

Practical applications

- To support the child's drive for autonomy, give children simple choices in order to reduce the likelihood of the child feeling confused.

- Setting clear boundaries and limits will aid the child to develop inner control (Garhart Mooney 2000:47).

Check! Learning Goals

You should be able to:

- name Erikson's first two stages and describe them briefly in your own words
- give two examples of how we can promote trust in the infant.

ATTACHMENT

Definition: 'a long-enduring, emotionally meaningful tie to a particular individual' (Schaffer 1996).

Attachment is recognised as one of the most important concepts in psychology and as such has been heavily researched. Our ability to form secure and meaningful relationships, not just with our parents/caregivers but also with others later in life, is considered a touchstone for healthy and happy functioning. Difficulties in this realm can have immediate and long-term consequences for the individual. We will look at the following topics in relation to attachment theory:

1. influences on attachment theory:
   - ethology
   - psychoanalysis
2. John Bowlby and maternal deprivation
3. Mary Ainsworth and attachment classifications.

Let's begin by tracing the history and development of attachment theory.

Development of Attachment Theory
Ethology

Ethology is the study of animal behaviour. Psychologists look to this area to gain insight into human behaviours.

Imprinting

This term was coined by Konrad Lorenz and it relates to the behaviour he observed in geese. When hatched goose chicks automatically attach to the first moving object they encounter. Lorenz found that when the geese hatched they attached to him even though he was not one of their species. Thus, Lorenz believed that 'imprinting' was instinctual and could not be reversed.

Harry Harlow

Harlow's study on the effects of maternal deprivation on rhesus monkeys consisted of removing newborn monkeys from their mothers and raising them in isolation without contact with others. The baby monkeys had access to two 'mothers' in their cage. One was made of wire mesh and had a bottle attached so the monkey could feed from it. The other was a 'cloth' figure covered in soft, tactile material but offered no opportunity for the monkey to feed. Harlow noted that when the baby monkeys where given the choice between a mesh wire figure with a feeding bottle or a figure covered in soft material, the monkeys chose to cling to the soft material rather than the other figure.

It was previously believed that the motivation for mammals to attach to a parent was for the purpose of feeding. However, Harlow's study indicated that monkeys sought 'comfort contact' and that social reasons were as important as, if not more important than feeding. Interestingly, when the monkeys were re-introduced to the troupe they were frightened, exhibited anxious behaviours and had difficulty socialising with other monkeys. This suggested that their early experiences created difficulties for them later in their social relationships with others.

Psychoanalysis

Mother–Child Relationship

Freud stressed the importance of the mother–child relationship. Many psychoanalysts believed that early experiences could affect a person in later life.

Object Relations

The famous psychoanalyst Melanie Klein developed the 'object relations' approach, which suggested that the 'loss' of an object could have a potentially negative effect. For example, to an infant an object could be their mother, and the loss of the mother, perhaps through death or separation, could in turn lead to negative outcomes for the child in later life.

John Bowlby

John Bowlby (1907–90) originally devised the basic tenets of this theory. Bowlby was working in a home for children with behavioural problems. These children had disrupted relationships with their families, and it struck Bowlby that these difficulties might hold the key to their emotional and behavioural disturbances. His time there prompted him to train as a child psychiatrist. He trained at the British Psychoanalytic Institute (BPS), where he was supervised by the prominent psychoanalyst Melanie Klein.

Bowlby agreed with Klein's view that early experiences shaped an individual in later life. He also supported Klein's 'object relations' approach and the potentially negative effect of the loss of an 'object'. However, unlike Klein, Bowlby believed that *actual* family experiences were important and could cause emotional problems in children. Bowlby also came to believe that relationship problems between parent and child continued to be passed down through the generations: thus, in order to help the child, a practitioner should look at and help the parent.

Bowlby was heavily influenced by the work of ethologists such as Konrad Lorenz and Harry Harlow (see above), whose work led Bowlby to believe that there was an instinctual aspect to the development of a bond between mother and child, and further that there was a 'critical' period for the child's development. We have seen that Harlow's research suggested that feeding alone, or the satisfaction of hunger, were not necessarily the primary motivation for bonding and that more social needs were at play. This research came to shape Bowlby's attachment theory.

Maternal Deprivation

Bowlby came to believe that if a child experienced a disruption in their relationship with their mother (through separation or death) this had a negative impact on the child. Bowlby believed that there was one fundamental attachment relationship and that was between mother and child. He termed this relationship 'monotropism'. He also believed that there is a critical period, between six months and three years, during which attachment must be maintained.

Criticisms of Bowlby

1. **Monotropism.** Research indicates that infants form several important attachment relationships with figures other than their mother (e.g. fathers, grandparents and siblings).
2. **Negative outcomes**. Bowlby emphasised that if a child experienced difficulties in their attachment relationship this could have disastrous outcomes in later life. Michael Rutter has rejected this bleak view, pointing out that all children experience separation at some point (e.g. going to school) and that we need to differentiate between different situations.
3. **Feminists** have attacked Bowlby's insistence that the mother must remain at home with the children to prevent the child suffering due to her absence. There have been suggestions that the timing of Bowlby's research is significant. Male soldiers were returning home after World War Two and the government hoped to encourage women to give up their jobs to the returning soldiers. Suggesting that children would be damaged by their mother's absence would discourage women from working outside the home.

Mary Ainsworth

Mary Ainsworth joined Bowlby's research team when she came to Britain from the USA. Whereas Bowlby formulated ideas about the nature of the attachment relationship, Ainsworth developed a way of testing the attachment relationship. She conducted a huge number of 'naturalistic' observations in Uganda and the US examining the behaviour of mothers and their infants. From these observations she developed a way of testing the quality of the attachment relationship through the 'strange situation'.

The strange situation consists of placing a mother and her child in a room where their behaviour can be observed. The child should use the mother as a secure 'base' from which to explore. During the experiment a 'stranger' will enter the room twice: once when the mother is present and once when the baby has been left alone. The baby's reaction to the return of the mother is used to gauge what attachment pattern exists. The reactions of the infants form the classification of attachment styles.

Attachment Classifications

Type A — Insecure/Avoidant

Babies exhibited an avoidance of interactions with the mother on her return. The baby either completely ignores the mother or else displays avoidance behaviours such as turning away or avoiding eye contact. During separation the baby does not display distress, or else the distress seems related to being left alone rather than to the mother's absence.

Type B — Secure

Babies classified as securely attached actively seek interaction and contact with the mother, especially during the reunion episode. If the baby shows distress during the separation episode this is judged to be solely related to the absence of the mother.

Type C — Ambivalent/Resistant

These babies were extremely upset when the mother left. On reunion with the mother the baby seemed to want to be near her yet 'resisted' her efforts to comfort them. If the mother picked them up they displayed a great deal of angry behaviours and tried to struggle free.

Type D — Disorganised

This classification was added later. It relates to babies who displayed 'disorganised' or disoriented patterns of behaviour that could not be classified under the other categories.

Behaviours that Promote Attachment

Bowlby believed that 'parental sensitivity' was important for the development of attachment. Ainsworth *et al.* (1978:152) found that these four scales were strongly linked to secure attachment:

- sensitivity
- acceptance
- co-operation
- accessibility.

Interestingly, De Wolff & Van Ijzendoorn (1997) found that 'playing' was an important factor in promoting attachment.

Just as certain behaviours can promote attachment, others can damage it or are more likely to elicit an insecure attachment pattern. Radke-Yarrow *et al.* (1985) examined patterns of attachment in two- and three-year-old children of depressed and 'normal' mothers. They found that the children of mothers with major depression were more likely to have an insecure attachment pattern (type A or C).

Internal Working Model

Bowlby believed that the child represents its relationship with its mother internally, and it is thought that this model serves as a sort of template for future relationships. According to Smith *et al.* (2003:98) internal working models are '... described as cognitive structures embodying the memories of day-to-day interactions with the attachment figure. They may be "schemas" or "event scripts" that guide the child's interactions and the expectations and affective experiences associated with them.'

Check! Learning Goals

You should be able to:

- identify three influences on the development of attachment theory
- describe what is meant by 'maternal deprivation'
- discuss the criticisms of Bowlby's theory
- name the four attachment patterns and explain them
- outline behaviours that encourage secure attachment.

THE DEVELOPMENT OF 'SELF'

The development of self-awareness:

- 18 months old — children learn to recognise themselves
- two years old — children begin to express their emotional states. Then they must make the distinction between self and other.

Implicit Self-knowledge

Rochat (2001) charts the development of self-knowledge in infants from implicit self-knowledge towards explicit self knowledge.

There are two types of implicit self-knowledge. The first is perceptual in origin and relates to the development of knowledge about their own body through self-exploration and action on objects.

The origin of the other type is social and refers to the development of specific knowledge through interactions and reciprocation with others.

The development of a 'sense of self' can be witnessed through our use of language. Rochat (2001) contends that the use of 'me' refers to the early stage of implicit self-knowledge, while the use of 'I' is accomplished at the explicit stage of self-knowledge or when self-concept is formed.

Check! Learning Goals

You should be able to:

- identify at what age a child recognises him/herself
- list the two types of implicit self-knowledge.

Personality and Temperament

Temperament is a biologically based propensity for individuals to react emotionally and behaviourally to events in a certain way.

What kind of temperament had you as a baby? Were you a happy or easygoing baby or did you frequently throw tantrums? Temperament is believed to contribute towards the development of personality. With regard to infancy, theorists have tended to focus on whether an infant's temperament impacts on the parent–child interaction. Another area of interest is whether temperament is stable across a person's lifespan and if childhood temperament predicts adult outcomes.

Thomas and Chess (1977, 1986) asked parents to fill out reports of their babies' behaviour and devised a classification of three types of temperament:

- **Easy babies** (40 per cent of sample) were reported to have regular feeding and sleeping patterns and were not fussy eaters. They were playful and reacted well to new situations.
- **Difficult babies** (10 per cent of sample), according to parental responses, had fussy

eating habits, irregular sleeping and eating patterns. They were irritable, threw tantrums when frustrated and cried more than other babies.

- **Slow to warm up babies** (15 per cent of sample) had mildly negative reactions to new situations. Unlike difficult babies, whose tantrums might include spitting out food, these babies showed more passive resistance, for example, they would let the food drool out rather than spit it out.

Can you see any difficulties with this study? It's based solely on parental reports of the infant's behaviour and as such might lack objectivity. Yet the idea of infant temperament is an important one and researchers have continued to develop ways of measuring it. Newer approaches include observing directly the infant's behaviour and asking not just the parents but other adults to rate the child's behaviour. Buss *et al.* (1984) found that temperament is not that stable in infancy, meaning it can change as the child grows older. Yet others have found stability in some aspects of temperament.

Shyness is part of a more general temperament style called **behavioural inhibition** (Passer & Smith 2001:472). Kagan *et al.* (1988) examined inhibited behaviours (infants who were quiet, shy and would withdraw from unfamiliar people and objects) and uninhibited behaviours (infants who were more sociable, spontaneous and open to new experiences). They found that extremely inhibited or uninhibited temperament was predictive of either childhood shyness or sociability. For example, a highly uninhibited infant became a sociable and talkative seven-year-old, while an extremely inhibited infant developed into a shy and quiet child. Kagan's work suggests that there is some evidence of long-term stability in certain aspects of temperament.

Check! Learning Goals

You should be able to:

- identify the three temperament styles from Thomas and Chess's research and briefly describe each of them
- discuss whether or not temperament is stable across the lifespan, making reference to Buss and Kagan's research.

In Focus: Community Mothers Programme

According to the Irish College of Psychiatrists (2005:17) there is a clear and documented relationship between mental well-being and early experiences, especially in the realm of attachment relationships. Further, 'Interventions specifically targeting this age group [0–5 years] can have preventative/protective value and have been shown to be successful (e.g. the Community Mothers Programme and programmes for the prevention of antisocial behaviour in childhood and adolescence).'

The Community Mothers Programme (CMP) is an example of Bronfenbrenner's ecological approach in practice. Bronfenbrenner suggested that the individual is influenced by their environment, which includes their immediate family right up to government policy. The Community Mothers Programme is a community-based programme funded by the HSE that seeks to support new mothers who are perceived to be vulnerable. Experienced mothers from the community are given training and then visit the new mothers to offer them guidance. This shows how Bronfenbrenner's approach, incorporating many different sources of influence in the child's life, can be put into operation in the form of a practical support programme. Let's take a closer look at a piece written by Director of the Community Mothers Programme, Brenda Molloy.

Mothers Helping Mothers
Brenda Molloy
A unique programme in which experienced mothers help other first time mothers, mainly in disadvantaged areas, has proven to be a great success.

Today's pressures mean that all parents, and particularly those who live in areas of social stress and disadvantage, need support if they are to promote the health and development of their children, their families and the next generation.

It is now accepted that there is a link between childhood experiences and adult outcomes. Failure to provide good quality support in the early years of child-rearing means that a much higher level of resources may need to be invested by the health, social and education services in later years in order to address and overcome the many problems that may arise as a result.

In addition to professional services, families need support networks to promote a sense of belonging and connection to the community.

The Community Mothers Programme, run by the HSE, has evolved since 1980, first using public health nurses as visitors to families with newborn babies, and then training experienced mothers from the local community to visit families to provide necessary support. Programmes like the CMP can provide a source of support to the family and help in building social networks.

The CMP is a support programme for first-time and some second-time parents of children from birth to 24 months who live in mainly disadvantaged areas. This includes lone parents, teenage parents, Travellers, asylum-seekers and refugees.

The CMP evolved from a UK-based child development programme. Following pilot phases, the programme was formally launched in the former Eastern Health Board in 1988. Today it is delivered to nearly 1,200 parents each year in the HSE Dublin/North East, and Dublin/Mid-Leinster regions.

The programme aims to support and aid the development of parenting skills, and enhance parents' confidence and self-esteem. It is delivered by non-professional volunteer mothers known as 'community mothers', who are recruited, trained and supported by family development nurses.

A key element in recruiting community mothers is that they reflect the ethos of the community they intend to visit. Each full-time family development nurse works with a team of 18–20 community mothers and supports 100–120 families at any one time.

The community mothers visit parents once a month in their own homes, providing empathy and information in a non-directive way to foster parenting skills and parental self-esteem. They use a clear and flexible set of strategies and focus on healthcare, nutrition and overall child development.

The community mothers are all experienced mothers who work on a voluntary basis. They are given nominal expenses for each visit. They typically spend upwards of 13 hours each month on their visits to between five and 15 families.

Community mothers' motivation is to help their community with the knowledge and experience each has gained through child-rearing. Participation in the programme helps to increase their feelings of self-worth as they see parents developing an understanding of child development and they find themselves gaining status in their own community.

A recent study also showed that volunteering in the programme contributed to lifelong learning. At the same time, the parents are empowered to believe in their own capabilities and skills for parenting without becoming dependent on professionals.

The monthly visit to the family is the main focus of the programme. The issues discussed at each visit are tailored to the particular needs of the family. The approach is supportive of the parents' own ideas and recognises the parents' desire to do what is best for their child.

The main focus of the community mother's visit is to encourage new parents, both mothers and fathers, to set themselves targets for achievement during the month before the next visit, and to facilitate the development of the child, both physically and mentally. This is done by drawing out the parents' own potential rather than by giving advice and direction. The community mother uses illustrated information sheets to show both effective ways of achieving childrearing goals.

The information sheets provide an easy, non-threatening and relevant way of raising difficult issues and discussing them, and they are also easily understood because of their direct style. The philosophy of the programme is simple yet profound. The parent is acknowledged as the expert with their own child and the programme works to support the parents to help them achieve their own goals for their child's development.

The family development nurse is available to the community mother to discuss problems and developments in relation to the programme. Once a month, the family development nurse meets with the community mother to discuss the families being visited. The community mothers also meet as a group, along with the family development nurse each month for support and ongoing training. Additional supports in the form of breastfeeding support groups and parent and toddler groups have evolved over the years. They are facilitated by community mothers and they support an additional 600 parents each year. In 1990 the programme was evaluated and was found to have significant beneficial effects for mothers and children.

Children in the programme scored better in terms of immunisation, cognitive stimulation and nutrition, and their mothers scored better in terms of nutrition and self-esteem than those in not in the programme.

At that time the programme was only aimed at first-time parents during the first 12 months of the child's life; parents received a maximum of 12 visits, usually one a month lasting approximately one hour.

Further evaluation was conducted seven years later, when the children were aged eight. A major finding was the persistence of superior parenting skills among the programme families. Children whose mothers were in the CMP were more likely to have better nutritional intake, read books and to visit the library regularly.

Mothers in the programme had higher levels of self-esteem. They were also more likely to oppose smacking, to have developed strategies to help them and their children to deal with conflict, to enjoy participating in their children's games, eat appropriate foods, and to express positive feelings about motherhood.

The benefits extended to subsequent children who were more likely to have completed their primary and MMR immunisation and to be breastfed. The results are positive as they show that just 12 contact hours in the first year of a child's life can make a difference.

The CMP programme empowers the women who deliver it as well as the parents who receive it. It is helping some 2,000 children a year to reach their full potential.

(Brenda Molloy is Director of the Community Mothers Programme, HSE Dublin/North East and Dublin/Mid-Leinster.)

Taken from www.irishhealth.com

CHAPTER SUMMARY

Hint. Always try to explain ideas in your own words. I test my understanding of a concept by trying to explain it to an imaginary person. (I wouldn't recommend doing this in public!)

Another way of making sense of and remembering all these concepts and theories is to apply them in some way to your daily life. Think of a young child you know: next time you see that child getting upset when a stranger goes to hold them, remind yourself that this is related to attachment theory, and so on.

For example, recently my friend's baby daughter was repeatedly kicking the mobile above her cot and it reminded me how amazing it is that she realises that she is separate from the environment and that she can make things happen (sense of agency) and that she was using movement (kicking) to learn more about her environment (Piaget's sensori-motor stage).

Physical Development

- The **cephalocaudal** principle of growth, i.e. from head to toe, is followed during this phase of growth.
- The **proximodistal** principle refers to development that goes from the centre out to the limbs and the extremities (hands and feet).
- The **brain** at birth is 25 per cent of its eventual adult weight; by the child's second birthday the brain will have increased to 75 per cent of its adult weight.

- **Sensation** and **perception** are associated with each other. Sensation includes vision, smell, hearing, taste and touch.
- **Visual perception**: Gibson's 'visual cliff' experiment — most babies refused to crawl over the edge.
- **Face perception**: Fantz found that infants showed a preference for the face that most resembled a human one.

Cognitive Development
Piaget's Sensori-motor Stage (0–2 Years)
Learning is through senses and movement (motor development).

As they develop, infants realise that they can make things happen: they bang spoons and throw objects from their high chair. This also highlights that the infant sees themselves as separate from their environment and that they are developing a sense of agency about their actions, in other words that they are the 'agent' or initiator of their actions.

Object permanence:
- refers to the understanding that objects continue to exist even when we can no longer see them
- its acquisition towards the end of this stage is considered an important milestone.

Imitation:
- Within a few weeks of birth babies are able to imitate a facial expression made by an adult.
- Meltzoff (1988) found evidence of what he believes to be a nine-month-old's capacity for 'deferred imitation'.
- Meltzoff and Moore's research would suggest that infant acquire concepts such as object permanence and mental representation earlier than proposed by Piaget.

Socio-emotional Development
Freud – Oral Stage (0–2 Years)
- Mouth is prime source of pleasure, feeding and weaning.
- Oral dissatisfaction caused by being weaned too early results in the infant becoming 'fixated' at the oral stage. This has negative consequences in later life.
- Orally fixated behaviours include swearing, smoking, nail biting, overeating, etc.
- Orally fixated personality is pessimistic, aggressive and distrusting.

Erikson

Trust vs. Mistrust (0–1 Year)

- How well the infant's needs are met and how sensitive the parenting is decides whether the infant develops trust or mistrust of the world.
- Encourage trust by: holding babies close; responding to them immediately when they are crying.

Autonomy vs. Shame and Doubt (1–2 Years)

- Autonomy means children can come to believe in themselves and their abilities.
- If the child's efforts are ridiculed or belittled they will develop a feeling of shame and doubt.
- Encourage autonomy by giving children simple choices and setting clear boundaries and limits.

Attachment

Definition: 'a long-enduring, emotionally meaningful tie to a particular individual' (Schaffer 1996).

Influences on Attachment Theory

Ethology

The study of animal behaviour.

1. Imprinting refers to the instinctual drive to attach to another. Witnessed in geese studied by Konrad Lorenz.
2. Harry Harlow's experiment: newborn monkeys were taken from their mothers and reared in isolation from others. 'Comfort contact' was important to the isolated monkeys who showed a preference for a cloth-covered wire mesh monkey over a mesh figure with a feeding bottle. When monkeys were re-introduced into the troupe they experienced difficulties mixing with other monkeys and exhibited anxiety.

Psychoanalysis

- 'Mother–child relationship' considered of vital importance. Early experiences affect later life.
- 'Object relations' (Melanie Klein): the 'loss' of an object could have a negative effect. A mother could be classed as an 'object' by the child and her 'loss' has the potential to negatively affect the child.

John Bowlby: 'Maternal Deprivation'

A disruption in a child's relationship with their mother (through separation or death) has a negative long-lasting effect on the child. 'Monotropism' refers to Bowlby's belief in one fundamentally important attachment relationship, usually with the mother.

There is a critical period (six months to three years) when attachment must be maintained.

Criticisms:

- Monotropism: children can form important attachments with more than one figure (e.g. father, siblings, grandparents, carers).
- Negative outcomes: Bowlby was pessimistic regarding the long-term effects of maternal deprivation on children. He did not believe that these effects were reversible. Rutter would disagree with that viewpoint.

Mary Ainsworth

'The strange situation': an experiment devised by Ainsworth to assess the attachment pattern of the child based on its reaction to its mother during the experiment.

Four attachment styles:

- Type A — Insecure/Avoidant
- Type B — Secure
- Type C — Ambivalent/Resistant
- Type D — Insecure/Disorganised.

Behaviours that affect attachment:

1 positively —
   - parental sensitivity
   - acceptance
   - co-operation
   - accessibility
   - play.
2 negatively —
   - major depression in mother.

The Development of 'Self'

The development of self-awareness:

- 18 months old — children learn to recognise themselves

- two years old — children begin to express their emotional states.

Rochat charts the development of self-knowledge in infants from implicit self-knowledge towards explicit self-knowledge.

There are two types of implicit self-knowledge: perceptual and social. Perceptual relates to the development of knowledge about their own body through self-exploration and action on objects.

The role of language in self-knowledge:

- the use of 'me' appears in the early stage of implicit self-knowledge
- the use of 'I' is accomplished at the explicit stage of self-knowledge.

Personality and Temperament

Temperament is a biologically based propensity for individuals to react emotionally and behaviourally to events in a certain way.

Thomas and Chess (1977, 1986) found three types of temperament:

- easy babies (40 per cent of sample) — regular feeding and sleeping patterns; playful; react well to new situations
- difficult babies (10 per cent) — fussy eating habits, irregular sleeping patterns; irritable; threw tantrums when frustrated
- slow to warm up babies (15 per cent) — mildly negative reactions to new situations.

Is temperament stable across the lifespan?

- Buss *et al.* (1984) — temperament is not that stable in infancy.
- Kagan *et al.* (1988) — extremely inhibited or uninhibited temperament is predictive of either childhood shyness or sociability.

CHAPTER SIX

EARLY CHILDHOOD: TWO TO FIVE YEARS

This chapter covers the physical, cognitive and social domains of development. Specific learning goals will be given at the beginning of each domain. There are two 'In Focus' topics in this chapter, the first dealing with the concept of 'prosocial behaviour' which will be examined in depth to give the reader a fuller picture of the various approaches used to gain insight into the development of this behaviour.

The second topic that will be addressed is autism: the causes, characteristics and treatment of this disorder.

CHAPTER OUTLINE

- Physical development:
  - Patterns of growth
  - Accidents
- Cognitive development:
  - Piaget's pre-operational stage
  - Vygotsky
  - Theory of mind
- Language development:
  - Is language uniquely human?
- Socio-emotional development:
  - Erikson's initiative vs. guilt (3–5 years)
- Freud's psychosexual theory
- Diana Baumrind's parenting styles
- Development of gender identity:
  - Sex differences
  - Gender stability
  - Gender constancy
- In focus:
  - Prosocial behaviour
  - Autism
- Chapter summary

PHYSICAL DEVELOPMENT

Patterns of Growth

From the ages of two to eleven years, the child experiences **synchronous growth**, which is a slow steady pattern of development. During this period the child is still growing in height and weight.

- height gain: 6.5cm per year
- weight gain: 2.25–3.2kg per year.

There are gender differences in growth: girls are generally lighter, smaller and have more fatty tissue; boys have more muscle tissue.

Growth between the ages of two and six years slows down a little, with gains of approximately 5–8cm a year in height and about 2.5kg in weight annually.

Changes to the brain:

- The number and size of nerve endings increase.
- The process of myelinisation continues during this period with a resultant increase in the speed of information transmission between cells.
- There is rapid growth in the frontal lobes.

Gross and fine motor skills:

- Gross skills seen during this period are hopping, jumping and running. As the child continues to develop they become increasingly organised in their motor abilities.
- Fine motor skills include better control of hands and arms as the body continues to develop. For example, a two-year-old child might be clumsy in their attempts to grasp a crayon whereas a five-year-old will exhibit far better manipulation and control of an object.

Accidents

According to the Children's Rights Alliance, in 2002 accidents were the leading cause of death in children. Furthermore, nearly half of all child injury deaths in Ireland are now caused by road traffic accidents, reflecting an upward trend since 1994. More age-specific statistics collected by the Western Health Board (WHB) in 1998 regarding injury admissions and the most frequent external causes in under fives identified the following:

- 40% were due to falls
- 21% were due to poisoning
- 7.8% were due to burns
- 2.8% were due to motor vehicle traffic accidents.

Figure 6.1: Developmental milestones

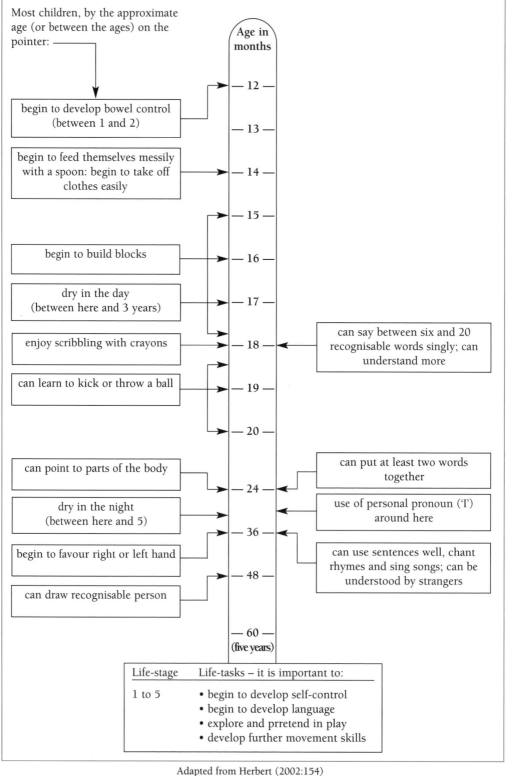

Adapted from Herbert (2002:154)

COGNITIVE DEVELOPMENT

Piaget's Pre-operational Stage (2–7 Years)

One of the most notable features of this stage is that the child is **egocentric**. This signifies that the child thinks that everyone else sees the world through their eyes. They do not understand that other people might see or think about things differently, and this is reflected in their thinking.

Symbolic thinking develops as the child begins to acquire language (which is of course made up of symbols), and they can use this newly developed ability to enhance their thinking. The child can now use words and images (symbols) to represent objects; for example, when a toddler sees a dog they might exclaim 'woof woof', using the noise a dog makes to represent a dog. By the end of this stage one would expect to see features such as animism begin to fade. Animism, as we saw in Chapter 2, is when the child ascribes consciousness to an inanimate object. An example of this is when the child hits their head on a table and then admonishes the table for being so naughty.

Piaget discussed **pretend play** as another feature of this stage: when a stick is transformed into a sword, for example. Piaget believed that play and **imitation** were important activities for the developing child. While he saw play's primary function for children as one of enjoyment he felt that imitation reflected the child's attempt to master or copy some new movement or action from another. Piaget proposed that while play involves a process of assimilation by the child, imitation is a product of accommodation in their construction of knowledge.

Piaget conceived that the development of play was broken into three parts. The second stage — the 'play stage' — is witnessed during the pre-operational period between the ages of three and six years old. This stage is dominated by the child's egocentrism and the child engages in mainly solitary play as they are generally too egocentric to play co-operatively with others. Parallel play does occur. In the pre-operational stage symbolic thinking is emerging and this is reflected in the child's use of objects and actions symbolically during their pretend play. (A cardboard box becomes a boat, for example.)

Practical Applications

Garhart Mooney (2000) suggests that during this stage:

1 children should be given large blocks of free play time
2 adults should provide 'real world' experiences for children — instead of looking at a picture of an animal it would be more stimulating for the child to have contact with that animal as it encourages a greater learning opportunity.

Check! Learning Goals

You should be able to:

- describe 'egocentric thinking'
- outline 'symbolic thinking' and its relationship to language
- explain 'pretend play'
- describe 'imitation' and explain why it is important to the development of learning in children
- describe the features of Piaget's 'play stage'.

Vygotsky

Vygotsky, another cognitive theorist, also emphasised the importance of play. Vygotsky believed that play allowed the child the opportunity to interact with other children, which in turn increased the potential for learning to occur. Another aspect of play, according to Vygotsky, is that language is used in the interaction between peers, through negotiation of rules for games and discussions around role playing. Vygotsky emphasised that language and development build upon each other.

Theory of Mind

Definition: Theory of mind is the understanding that others see and understand things differently from you.

The term 'theory of mind' is the ability to understand what another might be thinking. This ability is vital in the arena of social relationships: it enables us to predict others and have insight into their behaviours and actions. Piaget suggested that children do not develop this ability until the age of six or seven, but research has demonstrated that children typically acquire theory of mind earlier than this. The 'Sally Anne doll test' can be used to assess if theory of mind is present.

Imagine a young child watching a puppet show with two dolls named Sally and Anne. Sally decides to go for a walk but before she leaves she places her marble in her basket. After Sally leaves Anne removes the marble from the basket and places it in her box. **When Sally returns, where will she look for her marble?**

If the child replies that Sally will look for the marble in Anne's box (where Anne hid it) the child doesn't yet possess theory of mind. (How can Sally know the marble was moved there? The whereabouts of the marble is only known to the child who witnessed it being moved.) To demonstrate theory of mind the child has to be able to appreciate that Sally has her own beliefs about the world, which can differ from the child's own.

A deficit in theory of mind has been implicated in autism (see In Focus below for further discussion of autism).

Theory of mind is important in the social sphere of our lives, where the ability to understand other peoples' states of mind is essential.

Figure 6.2: The Sally Anne Doll Test

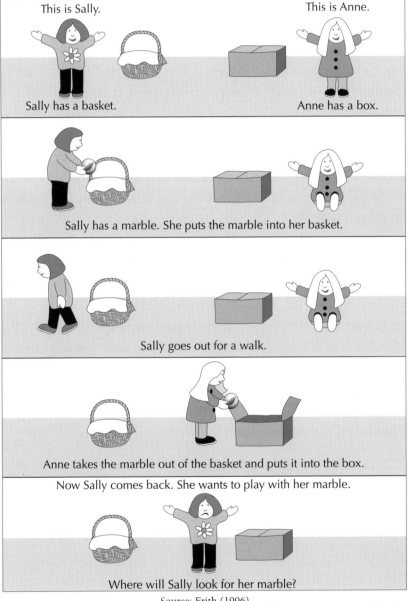

Source: Frith (1996)

Check! Learning Goals

You should be able to:

- define 'theory of mind'
- explain the 'Sally Anne doll test' and its purpose.

LANGUAGE DEVELOPMENT

Is language uniquely human?

The acquisition of language is undoubtedly one of humankind's greatest cognitive feats. Noam Chomsky and others such as P. Lieberman have argued that unlike any other species, including non-human primates, humans possess an inherited biological readiness to recognise and produce the structure and sounds of whatever language they are exposed to.

Yet research in the field of primates and language acquisition continues to grow and provide interesting results that continue to fuel the debate as to whether animals really can acquire language. The case of Washoe, a chimpanzee who was 'cross-fostered' — raised in a home and treated like a human child — spearheaded this debate. In 1966 Allen and Beatrice Gardner taught American sign language to the ten-month-old chimp. By the age of five Washoe had acquired 160 signs and was also able to combine signs. The Gardners replicated their research with four other chimpanzees, with the same result. It was also noted that the chimps started to sign to one another.

In light of this development Herbert Terrence also attempted to teach a chimpanzee sign language in the hope of facilitating communication between apes and humans in the future. He found his results disappointing and concluded that the chimp, named Nim Chimpsky, was not displaying language and was even unable to create a sentence. Terrence concluded that all that was being witnessed was the imitation of language. He did not believe that the primates were able to generate real language.

What do you think? How would you define language? Do you think Washoe understood sign language when she used it or do you accept Terrence's view that it was imitation?

SOCIO-EMOTIONAL DEVELOPMENT

Erikson's Initiative vs. Guilt (3–5 Years)

As the child becomes older they exhibit increasing curiosity and interest, they initiate play and question more. The main aim of this stage is to acquire a sense of purpose. If this is discouraged or they are held back they will not develop self-initiative and instead will hold back in later life.

Practical applications:

• encourage children to be as independent as possible

• focus on gains when children practise new skills, not on the mistakes they make (Garhart Mooney 2000:53).

Check! Learning Goals

You should be able to:

• explain this psychosocial stage in your own words

• outline the practical applications of this stage.

Freud's Psychosexual Theory

Anal Stage (2–3)

The focus of pleasure shifts to the anus, helping the child become aware of its bowels and how to control them. By deciding itself, the child takes an important step of independence, developing confidence and a sense of when to 'give things up'. Over-strictness about forcing a child to go or about timing or cleanliness can cause personality problems, such as anal fixation: forcing a child to go may cause reluctance about giving away *anything*. The person may become a hoarder or miser. Conversely, over-concern about 'going regularly' may cause obsessive time-keeping or always being late.

Phallic Stage (3–6):

Children become aware of their genitals and of sexual differences. This stage is when the paths of males and females begin to diverge as their differences become apparent to them.

The Oedipus Complex

Each boy unconsciously goes through a sequence of stages beginning with the development of a strong desire for his mother, noticing the bond between his parents (i.e. sleeping together), becoming jealous of his father and hating him, then becoming afraid of his father lest he discover his son's true feelings, which results in the final substage of fearing punishment — castration!

Check! Learning Goals

You should be able to:

- describe Freud's anal stage and what is meant by the tern 'anal fixation'
- explain the 'phallic' stage and the dominant feature of this period.

Diana Baumrind's Parenting Styles

Attachment is an important aspect of the study of the relationships between parent and child. Another area of research that has developed is parenting styles and the differences that appear to exist (e.g. some parents advocate a strict approach while others do not). Diana Baumrind in her paper 'Effects of authoritative parental control on child behaviour' (1966) proposed that there are three styles of parenting:

Permissive:

- parent tends to be relaxed about discipline, avoids exercise of control
- parent makes few demands on the child regarding household chores or orderly behaviour
- parent sees themselves not as an active agent shaping the child but as a resource for the child to use
- parent does not encourage the child to obey externally defined standards.

Authoritarian:

- parent values obedience, child in its place
- when there is disagreement between parent and child, parent favours a punitive approach
- parent believes the child should accept the word of the parent, does not believe in verbal give and take or discussions.

Authoritative:

- encourages verbal give and take
- willing to share reasoning behind their decision making with the child and is open to adaptation

- sets standards for future behaviour
- praises and affirms the child's qualities.

Implicit in this and other research on parenting styles is that different parenting styles can lead to different outcomes for the child. Dekovic and Janssens (1992) found that children of authoritative parents tended to be more popular, while children of authoritarian parents were less so. The relationship between parenting styles and prosocial behaviour in children is illustrated in the In Focus section at the end of this chapter.

The Royal College of Psychiatrists (2004) in its fact sheet on good parenting advocates the following strategies:
- be consistent
- give lots of praise
- plan ahead
- involve your child
- be calm
- be clear with your child
- be realistic.

Check! Learning Goals

You should be able to:
- list Baumrind's three parenting styles and briefly describe each
- list four strategies to promote good parenting.

DEVELOPMENT OF GENDER IDENTITY

Sex Differences

Toy Choice

From two years of age children appear to show sex differences in relation to their choice of toys. Boys tend to choose trucks, cars and blocks. They also engage more in play involving gross motor activity such as kicking and throwing a ball. Girls, on the other hand, tend to show a preference for dolls, dressing up and domestic play. In school, both opt for same-sex partners when playing, boys preferring outdoor play or sport-related activities whereas girls opt for indoor, sedentary activities.

Kuhn *et al.* (1978) found that by two and a half years of age children had beliefs about sex role stereotypes (a belief that certain activities are suitable for girls and vice versa). The young children reported that girls liked to play with dolls and to help their mother whereas boys liked helping their father.

Are you a Boy or a Girl?

This question will stump or confuse most two-year-olds. If you show them a stereotypical picture of a girl and a boy, however, 76 per cent will correctly identify their gender.

By 30 months this rises to 83 per cent and by 36 months 90 per cent answer correctly. So we see that as the child becomes older they are more likely to correctly identify gender and that younger children more easily identify their gender visually rather than verbally.

By three years old most children can correctly label their own and others' sex or gender. This marks the acquisition of 'gender identity'.

Gender Stability

According to Smith *et al.* (2003), at approximately four years old most children understand that the gender of a person remains the same or stable. If asked, a girl will answer that she'll be a mummy when she grows up.

Gender Constancy

By seven years of age, there is an awareness that being male or female (the biological sex of a person) does not change, for example if a boy grows his hair long, he is still a boy. As gender identity develops children also acquire **sex-role stereotypes**, beliefs regarding what is considered appropriate characteristics and behaviour for boys and girls. This process of learning is called **sex-typing**. As might be apparent sex-typing is learned through a process of **socialisation** in which children acquire the beliefs, behaviours and values of their group (society, culture).

Compare and contrast the cultural or social expectations regarding appropriate behaviour for a girl today compared to a girl born 50 years ago. Do you think attitudes and expectations have changed?

What do you think? Do you think the expectations of what is considered appropriate behaviour and sex roles in general are the same for females in the West compared to females in other parts of the world?

Check! Learning Goals

You should be able to:

- chart the development of gender identity
- indicate at what age children achieve gender constancy.

In Focus: Prosocial Behaviour in Children

Why do some people help others, at great risk to themselves and for no apparent benefit, while others display aggressive and antisocial behaviours? Are genes responsible, the family environment we are raised in, our peer group or perhaps our culture? Many researchers have grappled with these issues.

Prosocial behaviours have very practical consequences and implications for our understanding of human nature.

What is prosocial behaviour? The term refers to helping, caring, sharing, co-operation and sympathy (Hay 1994).

What processes or influences motivate prosocial behaviour? The empathy-altruism hypothesis was proposed by C. Daniel Batson. Empathy refers to the ability to put oneself in the place of another and to share what that person is experiencing, while altruism relates to the desire to help another without concern for oneself.

Hoffman's Empathy Development Model

Hoffman (1987) (cited in Schaffer 1996:271) described four stages in the development of empathy:

- **Stage 1: 'global empathy'** begins at the start of the first year and is characterised by the baby not seeing others as distinct and separate from themselves, so the child behaves as though what has happened to another has actually befallen them.
- **Stage 2: 'egocentric empathy'** starts in the second year. The child is now aware that it is a separate entity but continues to internalise the other's state.
- **Stage 3: 'empathy for another's feeling'** is found in children aged between two and three years. The child now recognises that others have distinct feelings.
- **Stage 4: 'empathy for another's life condition'**. This is the final stage in the development of empathy and covers early childhood. Here affect (feelings) is twinned with a mental representation of the other's general condition. Thus a concern for others in the form of an emotional response or a behaviour can be seen from the second year onwards.

Studies

Lois Murphy (1937), in her study of preschoolers and their interactions with peers, maintained that these young children demonstrated more instances of selfish and aggressive behaviour than of sharing and helping behaviours.

Piaget concluded that before the age of six children were not capable of caring about others.

These findings have been discredited by recent researchers:

Simner (1971) found that babies will cry in response to another baby crying.

Rheingold *et al.* (1976): children as young as eighteen months old have been found to share toys with another without prompting or encouragement. Rheingold's belief that prosocial behaviours are a 'natural' behaviour contradicts the suggestion of egocentrism. Zahn-Waxler *et al.* (1979, 1992) produced a series of findings that children's prosocial tendencies are most obvious when another is distressed. At the beginning of second year more attempts to comfort and ease suffering are witnessed in the form of patting and hugging, and by mid-second year verbal incidences of comforting attempts are both evident and increasing in frequency.

While prosocial behaviour increases in frequency from birth to two years, Hay (1994) has argued that between the ages of three and six there is a decline in prosocial behaviour.

However, Caplan and Hay (1989) observed that three- to five-year-olds were upset by another child's distress yet did not attempt to help or alleviate this distress. When the children were questioned as to why they had not helped they responded that there was an adult present and that it was not up to them to help. This example would support Eisenberg's (1989) assertion that a more complex approach is needed when viewing prosocial behaviours at this stage and points to the influence of gender and situation.

The Role of Genetics, Culture and Family

Genetics. While genetic influences can help explain acts of prosocial behaviour towards family members according to the *principle of kin selection* (Buck & Ginsberg 1991), this does not give an insight into examples of this behaviour towards non-kin. Sociobiologists have suggested the concept of *reciprocal altruism*, that one helps another in the belief that this will increase the likelihood of help being received (Trivers 1971).

This obviously does not satisfactorily explain the differences in prosocial behaviour.

Evidence has also been found in studies of identical twins and fraternal twins that points to greater similarity of the behaviour in the former in comparison with the latter (Rushton *et al.* 1986).

Cultural differences have been proposed as an explanation in the development and individualisation of prosocial behaviours.

Beatrice and John Whiting (1975) conducted studies of three- to ten-year-olds in six small communities in the Philippines, USA, Mexico, India, Kenya and Okinawa. They found that Kenyan, Philippine and Mexican children scored highest on altruistic behaviours and the American children scored lowest. A possible explanation for this result is that children from poorer backgrounds have more child-minding and other responsibilities as the mother often has to work in the fields. These children can witness the importance and genuine contribution altruism makes to their survival and that of their family. American culture tends to be more individualistic and the importance of altruism is not as pressing. Hindus were found to feel more obliged to behave prosocially than their American counterparts (Miller *et al.* 1990), which gives another aspect to the cultural influence.

Parenting: studies by **Zahn-Waxler** *et al.* (1979, 1992) propose that parental type is closely associated with children's prosocial behaviour.

These instrumental types are:

- *Provision of clear rules and principles*: if a parent explains a rule of behaviour the child is more likely to exhibit prosocial behaviour, for example explaining to a child why it should not bite another is more likely to elicit a positive response than merely telling them not to do it.
- *Attributing prosocial qualities to the child*: if a child is told that it is kind and so on then the child is more likely to internalise these characteristics as part of their own perceived personality.
- *Modelling by parent*: this is believed to be one of the most vital functions of parents and is almost common sense.
- *Empathic care giving to the child*: if parents are warm and responsive towards the child the likelihood increases of the child displaying the same tendencies.

What is clear is that there is no one formula that fits and explains prosocial behaviour entirely. The importance of peer group in the development of prosocial behaviours can be seen through programmes in schools that attempt to encourage this behaviour (Solomon *et al.* 1988).

Gender-wise it has been found that girls exhibit more kindness than boys, though this can be partially accounted for by socialisation and expectation. What is becoming clear is that prosocial behaviour tends to stabilise in mid-childhood.

In Focus: Autism

Autism lies along the pervasive development disorders spectrum. The very basic criterion for a diagnosis of autism is a qualitative impairment in reciprocal social interaction, in verbal and non-verbal communication and in imaginative activity. Individuals with autism can have mild to severe symptoms.

Autism was first described, independently of one another, by Hans Asperger and Leo Kanner in the 1940s. They chose to use the term autistic to reflect one of the main characteristics often witnessed: a withdrawal from others. *Autos* denotes 'self' in Greek and Frith (1996, p.7) comments that 'this narrowing could be described as a withdrawal from the fabric of social life into the self'.

According to the Irish College of Psychiatrists (2005:55), the estimated prevalence of this disorder is 60 in every 10,000 children and in Ireland it is estimated that there are approximately 5,330 children and adolescents (under the age of 16 years) with autism spectrum disorder. Further, 'it may coexist with intellectual disability or other disorders of development and can occur with other physical or psychological disorders. It is estimated that 40–60% of those diagnosed with autism spectrum disorder have intellectual disability.' It appears to affect males more than females: the male:female ratio for autism is 3 or 4:1.

According to the ICD-10 and the DSM IV:

- Autism is classified as one of the pervasive developmental disorders.
- It affects males predominantly.
- Individuals with autism display marked abnormalities in their capacity for reciprocal social interaction, in language and communication and in the development of symbolic play.
- With respect to communicative abilities a person with autism takes a literal meaning from what is said. If it was said that Mary had a hard neck, rather than believing Mary to be brazen the individual would believe Mary literally had a hard neck.
- People with autism also display repetitive behaviours and activities of play (Carr 2003).
- Individuals with autism also exhibit an apparent inability to empathise with others.

There is a spectrum of disorders, ranging from classic autism to high function autism, which is often termed Asperger's Syndrome. Like autism, Asperger's Syndrome is characterised by abnormalities in reciprocal interactions and restricted, repetitive patterns of activities and interests. However, it differs from autism in that no delay in language development or intellectual development occurs. Often people with Asperger's Syndrome have outstanding memories for facts or figures (Carr 2003).

Cognitive Development — Theory of Mind
Research conducted with children of normal development, with autism and with Down syndrome found that both groups of children without autism did well on the 'Sally Anne' exercise (which tests the existence of theory of mind), but the children with autism performed poorly. An example of this is a story related by Margaret Dewey (Frith 1996:182) of an autistic boy who was sent to the kitchen to get himself a drink of milk. His father came in to find the child pouring the carton of milk down the drain. The father started shouting at his son to stop, which the child did. He began crying but made no attempt to explain why he had done it. It later transpired that the child believed the milk to have gone off and was doing what he had witnessed his parents do with bad milk, pour it down the drain. The child did not defend his action when told to stop — exactly what one would expect from a child who does not realise that someone else does not necessarily have the same knowledge as himself. Thus being shouted at was a shocking turn of events that made no sense to him at all.

Biological Theories
Biological accounts stress a neurodevelopmental aspect and hold genetics as instrumental in the origins of autism. The autistic artist Jose, described in Oliver Sacks' book *The Man who Mistook his Wife for a Hat*, developed apparently normally until a childhood illness caused swelling to the brain that caused irreparable damage. Jose displayed many autistic characteristics and this could suggest that in some cases neurological damage can be held accountable.

Recent developments and research in neurology claim to have found that some autistic people have shorter brain stems than individuals with normal development. There is also a relatively recent association between the MMR immunisation vaccine and autism, which remains unproven.

Treatments
While there is no cure for autism (Cohen & Volkmar 1997), treatments do exist. These tend to take a behavioural modification approach; for example, the Lovaas method is popular, particularly for targeting challenging behaviours.

The TEACCH approach emphasises an intensive structured educative programme that aims to make the world intelligible to the autistic person by acknowledging deficits and building on the strengths of the child. The provision of a structured environment seems to be desirable, according to the findings of Rutter and Bartak (1973).

In Ireland, the Irish College of Psychiatrists in their paper 'A better future now' (2005) highlight the need for quick diagnosis and provision of service. They identify the following gaps in the service provision for autism:

- specialist autism services have developed unevenly
- specialist out-patient sector teams do not have resources to carry out initial assessments or to offer appropriate long-term interventions (p.56).

CHAPTER SUMMARY

Cognitive Development

Piaget

Features of pre-operational stage (2–7):

- egocentrism: the inability of young children to consider another person's point of view
- egocentrism comes from 'centration': the tendency to concentrate on one aspect of a situation to the exclusion of other viewpoints
- emergence of symbolic thinking
- animism: the child ascribes consciousness to inanimate objects. This should diminish towards the end of this stage
- pretend play and imitation are central to the child's learning.

Development of play. The second stage of play is witnessed in this period and is termed the *play stage*:

- dominated by the child's egocentrism
- mainly solitary play but parallel play can occur.

Vygotsky

Play is important as it allows for interaction between the child and its peers, thus increasing the potential for learning from others.

Theory of Mind

- Theory of mind is the understanding that others can have different thoughts and beliefs from you.

- Typically emerges at age six.
- Sally Anne Doll experiment can be used to assess whether TOM has been acquired.
- Its acquisition is considered fundamental to social interaction.
- A deficit in theory of mind has been linked to autism.

Socio-emotional Development

Erikson's 'Initiative vs. Guilt' (3–5 Years)

- The main aim of this stage is to acquire a sense of purpose. If this is discouraged or the child is held back they will not develop self-initiative.
- Encourage initiative by: encouraging children to be as independent as possible, focus on gains as children practise new skills, not on the mistakes they make.

Parenting Styles

Diana Baumrind's (1966) Parenting Styles

- Permissive — no limits placed on child, sees role as nurturer and resource for child to access.
- Authoritative — controls and boundaries put in place but willing to listen and negotiate.
- Authoritarian — very strict, rules are not open to change. Expects obedience.

See also Zahn-Waxler's parenting type associated with prosocial behaviour in children ('In Focus: Prosocial Behaviour in Children' section).

Development of Gender (Sex-Role) Identity

- Sex differences: toy choice. At two years of age, boys tend to choose trucks, cars, etc. Girls tend to show a preference for dolls, dressing up and domestic play.
- In school, both opt for same-sex partners when playing.
- Development of gender identity: as the child becomes older they are more likely to correctly identify gender.
- Gender identity: by three years old most can correctly label their own and others' sex or gender. This marks the acquisition of 'gender identity'.
- Gender stability: at about four years old, most children understand that the gender of a person remains the same or stable.
- Gender constancy: by seven years of age, awareness that being male or female does not change.

- Sex-role stereotypes: as gender identity develops children also acquire sex-role stereotypes — beliefs regarding what is considered appropriate characteristics and behaviour for boys and girls. The process of learning sex-role stereotypes is called 'sex-typing' and is learned through a process of socialisation in which children acquire the beliefs, behaviours and values of their group (society, culture).

CHAPTER SEVEN

MIDDLE CHILDHOOD: SIX TO ELEVEN YEARS

CHAPTER OUTLINE

- Physical development:
  - Growth patterns and general information
  - Motor development
  - Health behaviour in school-aged children
- Cognitive development:
  - Piaget's concrete operation stage (7–11)
  - Vygotsky — school and ZPD
  - Howard Gardner's theory of multiple intelligences
- Socio-emotional development:
  - Freud's latency stage (6–11)
- Erikson's industry vs. inferiority (6–12)
- Theories of moral development:
  - Piaget
  - Kohlberg
- The changing face of Irish families:
  - Irish children's and adolescents' concept of family
- In focus:
  - Children's understanding of well-being
  - Aggressive behaviour
- Chapter summary

For most children, this period in their lives is marked by some important milestones. They will have entered a whole new environment when they started school, marking another step towards independence. Peer relationships and learning in a new environment are challenges for the child. School can have a strong influence on the development of a child. For a child with a disability — from dyslexia to Down Syndrome — access and provision within the education system can be patchy and at times inadequate to meet their needs. Some children will experience bullying. Thankfully, though, for most it is a positive time. By the end of this stage they will be on the cusp of puberty, the transition from childhood to adulthood.

PHYSICAL DEVELOPMENT

Growth Patterns and General Information

This period is one of a **synchronous** pattern of growth, that is, slow, steady and consistent development. 'Baby fat' declines while muscle mass and strength improve. On average, children gain about 2.25–3kg in weight and approximately 5–7.5cm in height. Physical development during this time is viewed by some as the calm before the storm of puberty with the dramatic physical changes that come in its wake.

Motor Development

The increasing myelination of the central nervous system (brain and spinal cord) contributes to improved motor skill and co-ordination. Generally speaking there is a tendency for girls to be better co-ordinated in their movements while boys have greater speed and strength.

Table 7.1: Gross and fine motor skills from six to eleven years

Age	Gross Motor Skills	Fine Motor Skills
6	• Can catch and throw a ball • Can hop easily • Can skip in time to music	• Can build a tower of blocks that is straight • Can hold a pencil in a similar way to an adult
8–9	• Increased body strength • Faster reaction time • Can ride a two-wheeled bike easily	• Greater control over smalll muscles and therefore can draw and write with improved skill • Beginning to join lettters together in handwriting
10–11	• Boys and girls differ in physical marurity: girls experience puberty two years ahead of boys	• Greater ability for detailed tasks, e.g. needlework • Using joined-up letters

Health Behaviour in School-aged Children (HBSC)

This is part of an international research initiative examining health-related behaviours in children. In the following chapter we will study the findings of the HBSC Ireland surveys in greater depth. In the meantime we will note some of their findings on children's health.

Injuries

According to the Health Promotion Research Centre:

> Injuries and their consequences contribute to a silent epidemic experienced by young people throughout the world and are the largest cause of disability and death in children and adolescents in some countries. Injuries account for over 70% of all deaths in young people and the risk of injury rises dramatically as children enter adolescence.

A trend identified by the HBSC Ireland team between 1998 and 2002 is an increase from 40 per cent to 45.8 per cent in the percentage of children reporting they had been injured and who had been treated by a nurse or doctor. The increased trend is more evident in boys and the increase was witnessed across all three age groups ranging from ten to fifteen years old.

Nutrition, Exercise and Obesity

Obesity is becoming an increasing problem in the West, including Ireland. Research has indicated that an overweight child is more likely to become an overweight adult. According to Young-Hyman *et al.* (2006) increasing weight is associated with emotional and weight-related distress in children. Obesity is associated with poorer outcomes in general well-being, including the social, emotional and physical domains. The link between obesity, nutrition and exercise seems obvious. An increased concern has become noticeable in recent years with schools implementing bans on fizzy drinks and sweets as an emphasis is placed on the importance of good nutrition. Further, the role and importance of physical activity in children's daily lives has been highlighted.

Food Poverty

Another finding from HBSC Ireland relates to food poverty, which is defined as an inability to access an adequate and healthy diet. Sixteen per cent of young people surveyed reported going to bed hungry. These children are more likely: to come from a socially disadvantaged background; to have experienced extreme drunkenness; to smoke. They are less likely to report good health and feelings of happiness.

It is apparent that a relationship exists between good physical health and functioning and positive development in psychological areas, including self-esteem. From a lifespan perspective, children who are overweight when young are at increased risk of being overweight as adults.

COGNITIVE DEVELOPMENT

Piaget's Concrete Operation Stage (7–11)

This is the third stage in Piaget's theory of cognitive development. The main features of this stage are:

• conservation: the understanding that quantity, volume and length remain the same
• reversibility: the understanding that numbers or objects can be altered and then returned to their original state
• decentring: children become less egocentric.

Children learn through their interaction with concrete or 'real' objects, and a child can apply the learning strategies they have developed to real and immediate situations. However, the process of 'decentring', in which they become less egocentric, enables them to become more flexible in their thinking as they are now able to factor in other ways of looking or thinking about a situation. Another ability they develop during this stage is that of **reversibility**. This concept is best explained with an example. You have a ball of dough and you roll it into a long cylindrical snake-like object. Can you return it to its original state and would it still be the same? We know that it will stay the same because we possess the concept of reversibility, which is the understanding that we can alter an object and then reverse the process, returning the object to its original state. Reversibility is necessary for the acquisition of the concept of **conservation**.

Have you ever had to deal with warring children complaining that one is getting more drink than the other? You try to explain to them that even though the drink is in different shaped glasses they still have the exact same amount of liquid in each, though it might not look it. If they don't believe you it's probably because they haven't yet acquired the concept of **conservation**. As adults we can factor in the shape and size of the glass and understand that the amount of liquid is the same in both glasses even though it might appear not to be. During the concrete operational stage children begin to acquire this ability, and not just regarding the conservation of liquid/volume. Piaget believed that the acquisition of conservation was an important developmental milestone. The diagram shows that there are several forms of conservation.

Figure 7.1: Piaget's conservation studies

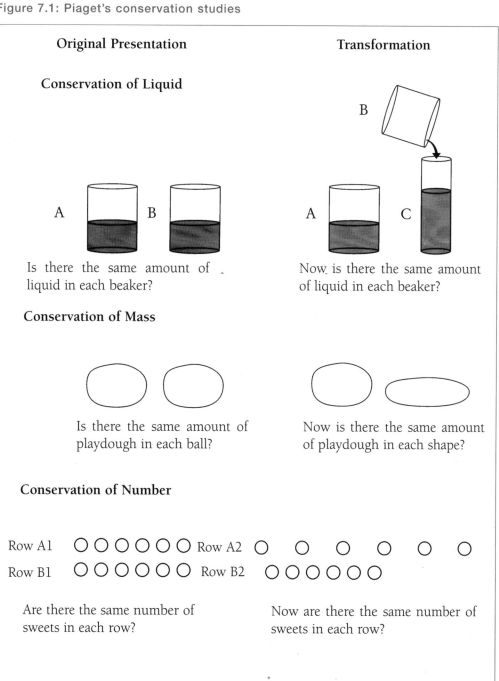

Original Presentation   **Transformation**

Conservation of Liquid

Is there the same amount of liquid in each beaker?

Now is there the same amount of liquid in each beaker?

Conservation of Mass

Is there the same amount of playdough in each ball?

Now is there the same amount of playdough in each shape?

Conservation of Number

Row A1 Row A2

Row B1 Row B2

Are there the same number of sweets in each row?

Now are there the same number of sweets in each row?

According to Smith *et al.* (2003), conservation is achieved at the following ages:

- conservation of number: about five to six years of age
- conservation of weight: seven or eight years of age
- conservation of volume: between ten and eleven years old.

Check! Learning Goals

You should be able to:

- describe decentring, reversibility and conservation
- indicate at what ages different types of conservation are achieved.

Vygotsky — School and ZPD

School marks an important milestone in the life of most children and it is here that you can see some of Vygotsky's concepts of how children learn.

Zone of proximal development (ZPD) refers to the distance between what a child can do or learn alone and what they can achieve with the help of others. For Vygotsky, teachers and more experienced peers act as a support to help the child bridge the distance between what they could learn alone and the further learning they could achieve with these supports. Vygotsky used the term 'scaffolding' to refer to the role played by teachers, who act as a bridge or scaffold enabling the child to achieve further learning, learning they could not as easily achieve on their own. Vygotsky believed that social interaction was essential to learning.

Practical suggestions:

- pair children together so they can learn from each other
- plan activities that are within the child's ZPD but that stretch them developmentally
- encourage conversations (Garhart Mooney 2000).

While Piaget believed that children learn through their own personal experiences, Vygotsky thought that they learn through their interactions with others.

Check! Learning Goals

You should be able to:

- explain the zone of proximal development
- describe how to encourage learning according to Vygotsky's theory.

Howard Gardner's Theory of Multiple Intelligences

> An intelligence is the ability to solve problems, or to create products, that are valued within one or more cultural settings (Gardner 1983).

In his book *Frames of Mind* Howard Gardner suggested that intelligence is not necessarily one unitary concept but rather several different types of intelligence that exist independently of one another but interact with each other all the time. Each of us has different strengths and these represent our unique mix of different types of intelligence. He identified seven types of intelligence:

- **linguistic** — reading, writing, talking, listening
- **logical-mathematical** — numbers, scientific thinking
- **bodily kinaesthetic** — using one's body, sport and dance
- **musical** — singing, playing, composing
- **interpersonal** — ability to understand and relate to others
- **intrapersonal** — self-understanding
- **spatial** — used in navigation, and parking your car.

Gardner later added two more types of intelligence:

- **naturalistic** — the ability to recognise and categorise objects in nature (plants and animals)
- **existential** — the capacity to ask questions regarding the meaning of life.

Gardner's theory has been heavily criticised on the grounds that it lacks scientific rigour and that some of his intelligences (such as inter- and intrapersonal) do not lie within the cognitive domains. Gardner has countered that criticism of his multiple intelligence approach represents a fear of moving away from measuring intelligence through the use of traditional standardised tests towards his less quantifiable approach in conceptualising intelligence.

Irish Education System

My main strengths lie in language and logical-mathematical intelligences, but I'm a lousy dancer with no sense of rhythm and I'm also tone deaf (lacking in kinaesthetic and musical intelligences). Therefore I have a unique mix: I have strengths in some intelligences and am less able in others. This of course raises questions about our educational system, which is mainly founded on two types of intelligence: linguistic and logical-mathematical. Under this current system I did well in school but if our education system had been founded on intelligences such as musical or bodily-kinesthetic, I would have been bottom of the class because my strengths do not lie in those areas.

> What do you think?
> What effect would it have on an individual who possesses musical intelligence if the school they attend has no resources devoted to music?
> What would be the school experience of someone who is not great at maths or linguistically based subjects?
> Do you think this could effect their self-esteem, make school more or less enjoyable for them? Could it put them at higher risk of truanting, perhaps?
> Should our school system change to encompass the variety of intelligences outlined by Gardner?
> If you are interested in this area visit the website of EMITT project St Brigid's School, Killarney, who have implemented Gardner's multiple intelligences theory into their curriculum: www.sip.ie/sip065/mitheory.htm.

Check! Learning Goals

You should be able to:

- list Gardner's seven types of intelligence
- describe each of these intelligences in your own words
- evaluate Gardner's different types of intelligence.

SOCIO-EMOTIONAL DEVELOPMENT

Freud's Latency Stage (6–11)

It's during this stage that sexual urges remain repressed and there are no significant developmental events. Freud believed that the most significant thing about children's social behaviour during this period was their preference to play and interact with same-sex peers.

Erikson's Industry vs. Inferiority (6–12)

This stage is very much influenced by the child's experiences in school. During this time the child begins to attend school and they interact more with their peers. If praised in their efforts they develop a sense of industry and feel good about what they have achieved, which encourages the feeling that they can fulfil their goals. If they repeatedly fail or are not praised when they try, a sense of inferiority may develop.

Check! Learning Goals

You should be able to:

- name Freud's and Erikson's stages and describe them in your own words
- explain the role of school in each theory.

THEORIES OF MORAL DEVELOPMENT

Schaffer states that:

> ... the end product of socialisation is an individual who can distinguish right from wrong and is prepared to act accordingly. Such an individual can be said to have acquired a sense of morality, that is he or she will behave in ways that uphold the social order and will do so through inner conviction and not because of a fear of punishment (1996:290).

Moral development can be seen as the understanding of 'right' and 'wrong'. The first thing that should be apparent to us is that different cultures/societies, even different families, can have very different ideas about what 'right' and 'wrong' are. Socialisation, as we saw in Chapter 6, is the process by which a person acquires the beliefs and values of the culture or society they live in.

Sociologists would contend that our moral development is socialised and we learn it from our parents (primary agents of socialisation) and then from school, church and our society in general (secondary agents).

Freud believed that children developed a moral conscience from their parents. If you think back to your childhood, no doubt you have memories of a parent or caregiver admonishing you for being bold or praising you for doing the 'right' thing.

Piaget

Piaget suggested that peer interaction at this age occurred around rule-based games. He developed a two-stage theory of moral development through his observations of young children playing and the way they used rules in play. Piaget noticed that younger children had less of a grasp of the rules than older children. Piaget questioned children of different ages about the rules of the games and formulated the following theory.

- **Pre-moral** — before the age of five, children do not consider what is right or wrong.
- **Moral realism** is witnessed in middle childhood: children's understanding of rules is governed by the influence of authority figures. They believe that rules can't be changed, that they are inflexible.
- **Moral relativism** is present from roughly eight years onwards. It reflects the developing child's growing sophistication in thinking. Children now understand that the rules can be modified through agreements with others. Rather than fixating on the rules being followed to the letter, they can modify the rules to meet their particular needs. They understand that this doesn't change the nature of the game. The important aspect is that all those participating are in agreement and following the same rules.

According to Piaget moral development was completed in children at approximately 12–13 years of age.

Piaget argues that children make the transition from moral reasoning and development-based rules imposed by adults or authority (constraint-based morality) to one based on 'mutual respect' where rules are negotiated from within rather than imposed from outside (Smith *et al.* 2003:273). Older children understand that rules are not absolute but are devices used by humans to get along co-operatively and therefore can be modified with the agreement of others.

> When, for example, the young child hears about one boy who broke 15 cups trying to help his mother and another boy who broke only one cup trying to steal cookies, the young child thinks that the first boy did worse. The child primarily considers the amount of damage — the consequences — whereas the older child is more likely to judge wrongness in terms of the motives underlying the act (Piaget 1932:137, cited in Crain 1985:118).

The Role of Piaget's Cognitive Theory

We have encountered Piaget as a cognitive theorist — someone interested in how children think and learn to think. Yet reasoning is a cognitive skill, so it's not such a jump to see that Piaget was also interested in how they reason on a moral level. Cognitive concepts such as **decentring,** the ability to see things from another point of view, play a role in the development of moral reasoning. Piaget contended that this ability is acquired by the age of ten, which may provide an insight into why Piaget used this age to mark the transition into his second stage of higher moral reasoning.

Criticisms of Piaget's Theory

One criticism of Piaget's conceptualisation of moral development is the idea that we complete this development at 12 or 13 years of age. It is recognised nowadays that development can constantly change across the lifespan and moral development is no exception. As we age, gain new life experiences and interact with more people, our attitudes change to reflect this. Thus one might anticipate our ideas of 'right' and 'wrong' may change.

Another criticism is culturally based. Ireland is a 'westernised' society which tends to be very individualistic. Other cultures are more 'collectivist' and 'traditional' in their views. In these societies great value is placed on traditions handed down through the generations and children are encouraged to embrace the traditions of their forefathers. Thus as children in these societies develop they become more 'constraint-based' in their moral reasoning, that is they look to and have greater respect for authority.

In the West, we are more individualistic, putting a high value on the rights of the individual. In more collectivist societies the individual is seen as part of a group and that group has a higher value than the needs of one particular individual.

Would you agree that a 13-year-old's ability to reason at a moral level is the same as an adult's?

Check! Learning Goals

You should be able to:

- explain what socialisation is and describe its role in moral development
- name and explain the three components of Piaget's theory of moral development
- outline criticisms of Piaget's theory of moral development.

Kohlberg

When we address Kohlberg's theory the role of thinking in moral reasoning will become apparent. Lawrence Kohlberg was heavily influenced by Piaget's work. Piaget proposed a two-stage approach to moral development. Before ten years of age a child believes rules are fixed, handed down by a higher authority (such as a parent or other adult) and not subject to change. After the age of ten, the child realises that rules are flexible and can be modified, with the agreement of others. Kohlberg went further than this and devised a six-stage approach. Kohlberg's theory is based on interviews he conducted with children and adolescents regarding moral reasoning. He devised 'moral dilemmas', the answers to which were used to categorise moral development.

One example of a moral dilemma is as follows.

A woman was near death from a special kind of cancer. There was one drug that her doctor thought might save her. The pharmacist in her town had recently discovered this drug and was charging ten times what the drug cost him to make. He paid $400 for the radium and charged $4,000 for a small dose of the drug. The sick woman's husband, Heinz, went to everyone he knew to borrow the money and tried every legal means, but he could only get together about $2,000, which is half of what it cost. He explained to the pharmacist that his wife was dying, and asked him to sell it cheaper or let him pay later. But the pharmacist said, 'No, I discovered the drug and I'm going to make money from it.' So Heinz gets desperate and breaks into the man's store to steal the drug for his wife. Should he have done it?

> What would your response be and why? Put this question to children and adolescents are varying ages and see what their responses are.

Using this dilemma, Kohlberg posed various questions, including whether it was right or wrong to steal the drug. The responses to these dilemmas form his theory of moral development. Kohlberg isn't particularly interested in whether the answer is 'yes' or 'no' but in the reasoning behind the answer.

Level 1: Preconventional Morality

Kohlberg thinks of this level as the one '… most children under 9, some adolescents, and many adolescent and adult criminal offenders' occupy (1976:33).

Stage 1: Obedience and Punishment

Whether something is 'right' or 'wrong' is determined by reward and punishment. There is no 'internalisation' of values: decisions regarding right or wrong are based on whether the result is rewarded or punished. For example, a child at this stage, if given the Heinz dilemma, would typically responded that Heinz was wrong to steal the drug because it's against the law. They see morality as something external to themselves, something decided upon by adults.

Stage 2: Individualism, Instrumental Purpose and Exchange
At this stage children recognise that there isn't just one correct viewpoint handed down by adults or the authorities. Each person has their own point of view. 'Heinz,' they might point out, 'might think it's right to steal the drug, the pharmacist would not.' Individual interests are an important aspect of this stage. One boy responded that Heinz might steal the drug if he wanted his wife to live, but that he doesn't have to if he wants to marry someone younger and better-looking (Kohlberg 1963:24), reflecting the notion of self-interest. They understand that there are no absolutes, that everything is relative. The concept of exchange in also seen during this stage, the idea of quid pro quo, or 'I'll do something for you if you do something for me.'

Level 2: Conventional Morality

This is the level most people acquire in our society. The main feature of this level is an understanding that norms and conventions are necessary to uphold society.

Stage 3: Mutual Interpersonal Expectations and Conformity
Living up to what is considered of you by others is considered important. Individuals value trust and loyalty as the basis for moral reasoning and judgements.

Stage 4: Social System Morality
Moral judgements are based on upholding the social values of the society you live in. Right is seen as contributing to the society, the group, etc. In this stage the individual sees themselves as a member of society and this influences their moral reasoning.

Level 3: Postconventional Morality

Kohlberg suggested that relatively few achieve these higher stages of moral reasoning, in which the individual, while valuing societal values and conventions, would go against them if a principle were at stake.

Stage 5: Social Contract
In this stage there is an understanding that differences can exist between a moral right

and a legal right. For example, if an individual helped a relation who was suffering to end their life, it would not be legally right yet some might feel it was morally acceptable. This example shows that there can be a difference between a legal and a moral right. Should a moral principle come into conflict with a societal value, an individual at this level would be willing to judge on the moral principle.

Stage 6: Universal Ethical Principles
Relatively few attain this stage of moral judgement, in which a person will follow their own principles of conscience.

Colby et al.

Colby *et al.* undertook a 20 year longitudinal study of moral judgement. Their findings included the following:

- the use of stages 1 and 2 decreased over time
- stage 4, which did not appear at all in the moral reasoning of ten-year-olds, was reflected in 62% of 36-year-olds.
- stage 5 did not appear until age 20–22, and never characterised more than 10% of the individuals studied (Colby *et al.* 1983).

The research suggests that Kohlberg's moral stages appeared somewhat later than he initially suggested. Further, the higher stages, in particular stage 6, were very elusive.

Criticisms of Kohlberg's Theory

A former student of Kohlberg, Carol Gilligan has been a major critic of his work. Gilligan claims that Kohlberg's theory is flawed because his research was based on male responses, so it represents a male perspective of moral development and reasoning. Gilligan, through her work with women, proposes that women apply an ethic of care in their judgements rather than an ethic of justice, which she claims is a male approach to moral judgements. An ethic of care means that the person considers other people and their welfare in their judgement-making process. Gilligan states that Kohlberg's work excludes and devalues women's approach to moral reasoning. Yet it is interesting that Gilligan falls foul of the precise accusation she levels at Kohlberg. Gilligan's research has been with females only, thus excluding and marginalising the male perspective.

Check! Learning Goals

You should be able to:

• describe Kohlberg's six stages of moral development

• outline Gilligan's criticism of Kohlberg's research.

THE CHANGING FACE OF IRISH FAMILIES

The Central Statistics Office (CSO 2006) reported that:

• in 1996 women represented 87.7 per cent of lone parents with children aged under 20; and this proportion increased to 91.6 per cent by 2006

• the number of women living as lone parents doubled from 60,100 to 115,000 over the period 1996–2006

• the number of men living as lone parents increased from 8,400 to 10,600 over the same period

• almost 85 per cent of female compared to over 96 per cent of male lone parents were aged 25 or over in 2006. The youngest child of 30 per cent of these women was aged between nought and four years

• just over two per cent of lone parents aged under 25 were male

• 80,366 persons were in receipt of one-parent family payments in 2005, of whom almost 98 per cent were women.

Ireland's family structure has changed in recent years from an almost universal traditional two-parent family pattern to a greater diversity in family forms. How do children perceive the diversity of families in Ireland? Let's look at some Irish research to find out.

Irish Children's and Adolescents' Concept of Family

Nixon *et al.* (2006) conducted research exploring Irish children's and adolescents' concepts of family. The authors note that '... childhood as an experience is socially, culturally and historically situated, and not simply a natural or universal state arising out of children's biological condition' (p. 79). In other words children are a part and product of their environment. Children also impact on their environments and shape them, so it is important to ascertain how children construe 'family'.

Ninety-nine children and adolescents were drawn from three schools in Dublin, their ages ranging from nine to sixteen years of age. The first part of the research involved a focus group to examine children's concept of family. Five vignettes, portraying different types of family, were presented to the children with the question, 'Are they a family?' The second part of the research involved the use of open-ended questions to explore children's definitions of family. The five vignettes were taken from O'Brien *et al*. 1996. Let's take a closer look at the vignettes used and the responses given by the children.

Vignette 1: 'John and Susan are a married couple without any children.'

- Almost two-thirds of the children endorsed this grouping as a family.

- The authors report that the most frequent reason given for not considering this grouping as a family was the absence of children. They believe this highlights the important role children attached to themselves within the family unit.

- 'If they have a dog they are ... it's like they're not part of a family yet because they need something to commit their life to' (12-year-old girl).

Vignette 2: 'Janet and Dave are a married couple with a son called Ben.'

- All the children endorsed this grouping as a family unit primarily because of the role of the members to care for each other.

- The authors relate that other reasons given for its endorsement as a family unit were: presence of a child; the parents are married; and the family members live together.

- 'They are a family because, like, they all take care of each other and they have — they have more responsibility to take care of more people' (ten-year-old girl).

Vignette 3: 'Jim and Sue live together with their six-year-old called Paul. They are not married.'

- Over 80 per cent of the children believed this grouping represented a family.

- One-eighth of the children indicated that they did not believe that this grouping constituted a family — this appears to be related to the absence of marriage and the children's perceived sense of insecurity and instability that entailed.

- 'Because like when them two have committed their lives to him [the son] but the mam and dad haven't committed their lives to each other ... the dad can walk out any day now but, like, if they're married, they have to stay together or get a divorce' (12-year-old girl).

Vignette 4: 'Sally is divorced with a ten-year-old daughter, Karin. Karin lives with Sally.'

- 90 per cent of the children responded that this grouping constituted a family.

- The main reason given was the emotional ties that connect family members. The authors surmise that 'children were able to distinguish between the inter-parent relationship and the parent-child relationship, and also understood that the quality of the two relationships were independent.'

- Six children were unsure if this grouping did represent a family. 'They are a family but a bit of the family has broken off , so it's not fully a family' (12-year-old girl).
- The authors continue, 'there was a sense that the divorced family was less than a proper family and represented a deviation from the ideal form'. The authors continue that the findings suggest that for a minority the concept of a nuclear two-parent family remains central.

Vignette 5: 'Karin's father, Tom, lives at the other end of town.'
- 80 per cent of the children endorsed this grouping as a family.
- Six children were uncertain, citing that the father and daughter did not live together.
- 'I don't think they are at all. I'd say they're just related, like when you don't live with one of your parents. It seems as if they're just somebody you go to see sometimes. It doesn't seem like they're your family at all' (15-year-old girl).
- The authors remark, '... the importance of the quality of family relationships emerged in the discussion of this vignette ... a family was dependent upon the levels of contact and the nature of the relationship ...'

The research also consisted of open-ended questions whose aim was to explore **how children define family** and also how children view their role and the parental role within the family. The research's main finding was that '... a family was defined in terms of the roles and relationships within the family'. Age differences were present, with younger children perceiving family as 'people who love you and take care of you, a group of people all born from the same person, and a group of people who live in the same house'. The adolescents used the following terms when defining family; 'closeness', 'a safe place to be' and 'people who comfort each other'.

Parental Role

For the children the parents' main role was to look after the children. Age differences were visible in how the parental role in the family was understood.

Younger children's perceptions of the mother's and father's role within the family were quite stereotypical: the mother was considered the primary caregiver and the father provided financial support. The children related that the father was more willing to take part in play activities than the mother.

Older children, while acknowledging the nurturing and 'breadwinner' roles, mentioned that they saw the role of both parents as one of 'protector' and 'confidant'.

Child's Role

The researchers found that '... overall children believed they had an active and important role to play in family life.' This included taking part in decision-making and caring roles such as looking after a sick parent.

The authors conclude that the '...main finding that emerged from their responses was that Irish children and adolescents were accepting of a variety of family forms: they conceptualised 'family' in an inclusive and flexible way.'

In Focus: Children's Understanding of Well-being

In Chapter 2 we looked at the development of the government's National Children's Strategy. Part of the strategy is the development of the National Child Well-being Indicators. The National Children's Office (NCO) took the following as its definition of well-being:

> Healthy and successful individual functioning (involving physiological, psychological and behavioural levels of organisation), positive social relationships (with family members, peers, adult caregivers, and community and societal institutions, for instance, school and faith and civic organisations), and a social ecology that provides safety (e.g., freedom from interpersonal violence, war and crime), human and civil rights, social justice and participation in civil society. (Andrews *et al.* 2002:103).

We shall look at some very innovative research carried out by Nic Gabhainn and Sixsmith (2005), whose aim was to explore Irish children's understanding of well-being. The research was innovative because of the method used by the researchers — they distributed disposable cameras to children of primary and secondary school age, from both rural and urban backgrounds. The children were asked to take pictures of things that represented well-being to them. The pictures were developed and another set of children sorted them into themes. Finally, older children were gathered to discuss the schemes that had been earlier agreed upon. Several findings emerged from the study regarding children's understanding of well-being.

Themes such as family, friends, school, music and sport emerged, in line with other research, but interestingly pets were identified in this Irish research as being important to children's well-being. The authors state:

> A number of specifically interesting factors emerged during both the categorisation and schema development phases. The centrality of interpersonal relationships with family and friends (including school friends) emerges strongly, as does the value of activities or things to do. During all phases, children discussed how aspects of the various categories made them feel; how relationships (with people and animals) and the activities within or context of those relationships gave them a sense of belonging, being safe, loved, valued and being cared for. In a sense, this reflects a returning to the original explanation of well-being that was provided to those involved in taking the original photographs: *feeling good, being happy and able to live your life to the full.* While many of the categories identified may have been predicted from reviewing previous work and so validate these aspects, the categories of pets/animals and environment/places are relatively unexpected yet credible as they were confirmed by the different groups of children independently at the three stages of the study and integrated in the schema development when the opportunity was available to reduce their importance. These categories illustrate the extent to which children interact with the natural world around them, as well as the interpersonal environment in which they are located. These factors certainly deserve further attention from researchers, policymakers and practitioners. (2005: 64)

Research should inform policy, which should inform practice. We can but hope that this research illustrates the interesting methods that can be used to obtain information; but also the importance of gathering research to gain further understanding of children's lives.

In Focus: Aggressive Behaviour

In Chapter 6 we focused on prosocial behaviour, exploring the different theoretical approaches to its understanding and also the possible influence of nature (biology, genetics) and nurture (culture). In this chapter we'll investigate the opposite side of prosocial behaviour: aggressive behaviour.

In recent times there appears to be an increasing opinion that our society is becoming more violent, and particular concern has been voiced regarding the behaviours of pre-teens and adolescents. Aggression is a behaviour: how does that behaviour develop and manifest? These questions have been considered by psychologists who have suggested various explanations. Let's have a closer look.

Defining Aggression

Defining aggression creates many difficulties, especially when dealing with *intentionality*. Is there a difference between a person who does harm when under the influence of drugs and a soldier on the battlefield? If a child pulls a toy away from another because they want to play with it, and the other child is hurt in the process, did the child intentionally mean to cause hurt?

In studies of aggression there appears to be a schism in how it is approached: whether aggression is defined by *intentionality* or by *result*.

Schaffer (1996:279) suggests dividing the categories of aggression into:

- *hostile aggression* — the intention of the act is to harm another
- *instrumental hostility* (as in the example of the child and the toy) — the action is aggressive but the motivation is non-aggressive.

A distinction between physical aggression and relational (relationships) aggression is also helpful, especially when dealing with gender difference. Research into relational aggression is helpful in explaining the differences in levels of aggression between boys and girls and perhaps presenting a more complete picture of female aggression. Thus far gender difference has been accredited to hormonal and socialisation differences.

In the past adult aggressive behaviours were studied without reference to childhood aggressive behaviours. It has now been proposed that aggressive behaviour in childhood can become stable and persist into adulthood.

- Facial expressions of anger have been observed in infants as young as three months old (Izard *et al.* 1995).
- Yet even at this young age gender differences have been reported, with boys more

likely to express anger than girls. As the infant becomes older an increase in verbal aggression is witnessed as the child is no longer confined to physical outbursts alone. This might explain the decrease in expressions of anger that has been reported in pre-school children.

• Hartup (1970) found a decrease in frequency of aggressive incidents in children between the ages of four and seven years of age, though as stated above this could be a reduction in instrumental aggression rather than in aggression per se.

• Aggression observed at six to ten years old correlates with aggression towards peers at ages 10 to 14 (Olweus 1991). With respect to stability, several studies have reported that children who are perceived or believed to be aggressive by a certain age in childhood are more at risk of continuing to be aggressive in adulthood.

Eron's Studies

One of the most influential studies relating to the stability of aggression was carried out by Eron, who sampled 600 eight-year-olds over 22 years (Eron *et al.* 1971). The children were first rated at eight years of age, based on peer perception and their own perception of their aggression. Researchers contacted the sample again (modal age 19) and managed to re-interview 427 of the original sample.

Eron reported that one of the most impressive findings was the stability of the aggression over time and also noted that their intellectual ability was negatively related to aggressive behaviour. Finally Eron and his team contacted the sample (modal age 30) and re-interviewed 295 in person and 114 by mail. They also interviewed the spouses and children of some of the sample. What they discovered was the continued stability of aggression but also that as parents, they were more likely to punish their children severely and be aggressive towards their spouses. Eron concluded that, 'by the time the child is 8 years old, characteristic ways of behaving aggressively or non aggressively have already been established.'

Farrington (1991) supports this finding, having undertaken similar research in England, and he found that, in males, aggressiveness in mid-childhood was an important predictor of antisocial activities in adulthood.

Theories

Many theories have been put forward in an attempt to shed light on the determinants of aggressive behaviour.

Temperament and emotional regulation. Children who are reported by their parents as having difficult temperaments are more likely to experience behavioural problems and aggression (Kingston & Prior 1995). Naturally the parent–child relationship has to be factored into the parents' perception that their child is difficult.

Cognitive. It has been argued that aggressive children have a greater inability to solve problems on a cognitive level and also to understand others' intentions and motives, resulting in an inappropriate aggressive response.

Ethological. Lorenz's ethological theory posits that humans are inherently or naturally aggressive and have learned ways to control aggressive tendencies. Lorenz emphasised the evolutionary value of aggressiveness with regard to survival.

Social-learning. Bandura argues that humans learn to aggress in his social-learning theory. Bandura's research has centred on children's imitation of aggression, particularly from those they admire. This argument has very practical implications with regard to violence on television, video games and films. Much of Bandura's research demonstrates the relationship between children observing and copying aggression. Yet interestingly, Schuck et al. (1971) found that after interaction with violent media girls showed a decrease in aggressive tendency whereas in boys an increase was noted.

Parenting and family environment. The family environment and relationships, as witnessed in Eron's findings, play an influential role in the development of aggressiveness. Several types of parenting have been associated with increased levels of aggressive in children.

Dan Olweus (cited in Schaffer 1996:287) identified several parenting styles which he proposed were implicated in children's aggression.

- 'Rejection by parents' — adolescent boys whose mothers are indifferent or fully reject them behave more aggressively. The mother's indifference is reflected in a lack of interest in the youth's attempt to develop self-control, and a lack of praise.
- 'Parental permissiveness' — when a parent does not set limits of behaviour and maintain them the boy feels he can behave as he wishes with apparent parental approval. Olweus found a strong relationship between a high incidence of aggression and a high degree of laxness on the mother's part.
- 'Parental modelling of aggression' — children imitate and learn behaviours, including aggression, from their parents. Aggression has been traced across three generations. (Eron study)

Culture. Cultural factors have a part to play in the development and maintenance of aggression. The Great Whale River Eskimos' emphasis on peace and their abhorrence of violence and aggressiveness is reflected in their child-rearing practices, in which they actively discourage aggressive behaviour (Honigmann 1954). In other cultures and subcultures, however, toughness and aggressiveness are behaviours often revered and encouraged.

> What do you think?
>
> Which of the theories put forward do you think provides the most compelling explanation for aggressive behaviour, and why?

CHAPTER SUMMARY

Physical Development

Synchronous pattern of growth — slow, steady and consistent development. On average, children gain 2.25–3kg in weight and approximately 5–7.5cm in height.

Injuries and their consequences are largest cause of disability and death in children and adolescents in some countries. HBSC Ireland reported an increase from 40 per cent to 45.8 per cent in the percentage of children reporting they had been injured and had been treated by a nurse or doctor in the period 1998–2002.

Obesity is becoming an increasing problem in the West, including Ireland, and is associated with emotional and weight-related distress in children.

Food poverty is defined as an inability to access an adequate and healthy diet. According to HBSC Ireland 16 per cent of young people surveyed reported going to bed hungry.

Cognitive Development

Piaget's Concrete Operation Stage (7–11)

Main features:

- **conservation** — the understanding that quantity, volume and length remain the same
- **reversibility** — numbers or objects can be altered and then returned to their original state
- **decentring** — children become less egocentric.

According to Smith *et al.* (2003), conservation is achieved at the following ages:

- conservation of number: about five to six years of age
- conservation of weight: seven or eight years of age
- conservation of volume: between ten and eleven years old.

Vygotsky – School and ZPD

Zone of Proximal Development refers to the distance between what a child can do or learn alone and what they can achieve with the help of others. A child's entrance into school allows for teachers to become a support to enhance children's learning

Vygotsky's term '**scaffolding**' refers to the role played by teachers in encouraging the child towards further learning, learning they could not as easily achieve on their own.

Piaget believed children learn through their own personal experiences, Vygotsky thought they learn through their interactions with others.

Howard Gardner's Theory of Multiple Intelligences

Howard Gardner identified seven types of intelligences in his book *Frames of Mind*:

* **linguistic** — reading, writing, talking, listening
* **logical-mathematical** — numbers, scientific thinking
* **bodily kinaesthetic** — using one's body, sport and dance
* **musical** — singing, playing, composing
* **interpersonal** — ability to understand and relate to others
* **intrapersonal** — self-understanding
* **spatial** — used in navigation, and parking your car.

Gardner later added two more types of intelligence:

* **naturalistic** — the ability to recognise and categorise objects in nature (plants and animals)
* **existential** — the capacity to ask questions regarding the meaning of life.

Gardner suggested that intelligence is not necessarily one unitary concept, but several different types of intelligence that exist independently of one another but constantly interact with each other.

Socio-emotional Development

Freud's Latency Stage (6–11)

Sexual urges remain repressed, children prefer to play and interact with same-sex peers.

Erikson's Industry vs. Inferiority (6–12)

Influenced by experiences in school.

* positive outcome — child develops a sense of industry and achievement
* negative outcome — a sense of inferiority may develop.

Theories of Moral Development

Definition: '... an individual who can distinguish right from wrong and is prepared to act accordingly. Such an individual can be said to have acquired a sense of morality, that is he or she will behave in ways that uphold the social order and will do so through inner conviction and not because of a fear of punishment' (Schaffer 1996).

Piaget

- **Pre-moral** — before the age of five, children do not consider what is right or wrong.
- **Moral realism** — occurs in middle childhood: children's understanding of rules influenced by authority figures. They believe that rules can't be changed, that they are inflexible.
- **Moral relativism** — present from roughly eight years onwards. Children now understand that the rules can be modified through agreements with others.

According to Piaget moral development was completed in children at approximately 12–13 years of age.

Kohlberg

After ten years old, the child realises that rules are flexible and can be modified with the agreement of others. Kohlberg devised 'moral dilemmas' (e.g. Heinz dilemma), the answers to which were used to categorise moral development.

1 Level 1: Preconvential Morality. This level is the one '... most children under nine, some adolescents, and many adolescent and adult criminal offenders' occupy.
   - Stage 1: Obedience and Punishment. Whether something is 'right' or 'wrong' is determined by reward and punishment. Morality seen as external, decided on by adults.
   - Stage 2: Individualism, Instrumental Purpose and Exchange. Children recognise that each person has their own point of view. They understand that there are no absolutes, everything is relative.
2 Level 2: Conventional Morality. This is acquired by most in our society. Main feature — understanding that norms and conventions are necessary to uphold society.
   - Stage 3: Mutual Interpersonal Expectations and Conformity. Individuals value trust and loyalty as the basis of moral reasoning and judgements.
   - Stage 4: Social System Morality. Moral judgements based on upholding social values of the society you live in.

3   Level 3: Postconventional Morality. Relatively few achieve these higher stages of moral reasoning. The individual values societal values and conventions but would go against them if a principle were at stake.

- Stage 5: Social Contract. An understanding that differences can exist between a moral right and a legal right.
- Stage 6: Universal Ethical Principles. Relatively few attain this stage of moral judgement — a person will follow their individual principles of conscience.

Criticisms of Kohlberg's Theory

Carol Gilligan claims that Kohlberg's theory is flawed because his research was based on male responses and thus represents a male perspective of moral development and reasoning. Gilligan's work with women led her to propose that women apply an ethic of care in their judgements rather than a 'male' ethic of justice.

The Changing Face of Irish Families

CSO (2006) reports that:

- the number of women living as lone parents doubled from 60,100 to 115,000 over the period 1996–2006
- the number of men living as lone parents increased from 8,400 to 10,600 over the same period.

CHAPTER EIGHT

ADOLESCENCE

We're just bored teenagers
Looking for love
Or should I say emotional rages
Bored teenagers
Seeing ourselves as strangers
(FROM 'Bored Teenagers', THE ADVERTS, 1978)

CHAPTER OUTLINE

- What is adolescence?
- Physical development:
  - General
  - Growth patterns
  - The adolescent brain
  - Sexual maturation
  - Timing of puberty
  - Health behaviour in school-aged children (HBSC)
- Cognitive development:
  - Piaget's formal operational stage (12 onwards)

- Adolescent egocentrism theory
- Socio-emotional development:
  - Erikson's fifth stage: identity vs. role confusion
  - Marcia's theories of identity
  - Self-concept and self-esteem
- In focus:
  - Eating Disorders
  - Suicide
- Chapter summary

Teenage angst is familiar to most of us. I still remember my teenage years and feel almost a sense of sadness for my son: the challenges he'll face, the intense self-consciousness and the many dangers I perceive to be ahead of him. Yet this is balanced by the knowledge that he faces an exhilarating journey with many positive things along the way. I remember feeling so strongly about things — the ideals I had and how I could put the world to rights.

During my teens I became a vegetarian, was passionately anti-vivisection (against animal testing to you and me) and brought a copy of the *Socialist Worker* home to my father to wind him up — it worked! I also recall feeling that no one really understood me and when I suffered my first heartbreak, despite the protestations of my family that they had experienced the same, I didn't believe it at all. I looked at these adults and thought there was no way they had felt as deeply as I did and that I would never recover from it. Of course I did, but this is just a small memory of what it is like to be a teenager.

WHAT IS ADOLESCENCE?

The word adolescence is derived from the Latin *adolescere*, which means 'to grow up' and is seen as a unique transitionary period from childhood to adulthood encompassing not just physical but also social, emotional and cognitive changes. **Puberty** refers to the commencement and maturation of biological, physical and sexual characteristics and in most countries is viewed as signifying the change or transition from childhood to adulthood. It refers to the physical and sexual changes that are witnessed during this period.

Adolescence encompasses the behavioural, social and emotional aspects: these are influenced by physical and social/cultural factors.

The World Health Organisation defines adolescence as the second decade of life, from ten to 20 years of age. Yet adolescence is not merely chronological age, it is also socially constructed, which means that some societies have different ideas about what 'adolescence' means. In some cultures puberty marks the child's coming of age and ability to take on 'adult' roles. In some countries girls marry or begin childbearing far earlier than we would find normal in the West.

PHYSICAL DEVELOPMENT

General

Looking first at the general physical development in adolescence and at puberty, the process of physiological change is set in motion with the release of hormones by the pituitary gland in the brain. The increase in these hormones begins at approximately eight or nine years of age. At this stage we also start to see gender differences emerge as different hormones effect different changes. In boys there is an increase in the hormone testosterone and in girls an increase in oestrogen. These hormones create physical and sexual changes.

- In boys: shoulders broaden; legs lengthen; they gain more muscle.
- In girls: hips widen; they gain fat; they become more 'curvy'

There is a difference in the age when males and females start their growth spurt:
- girls: approximately ten years of age
- boys: approximately 12 years of age.

Girls begin their development approximately two years earlier than boys. It is worth noting that these ages merely indicate the beginning of the biological changes, not necessarily the secondary sexual characteristics associated with adolescence, which we will look at later.

Girls tend to be better co-ordinated in their movements, while boys may have greater speed and strength.

Growth uses up fuel, so appetite and the need for sleep will increase to support growth. Levels of activity will also vary, as growing consumes energy. Fatigue from growth will also induce changes in mood such as irritability and lethargy. The response of parents and teachers at this stage should be to encourage the consumption of a balanced diet with adequate calories for the body and brain, and to recognise the need for sleep and rest.

Growth Patterns

Adolescence is a spectacular time of change for the human body. We encountered the proximodistal and cephalocaudal principles of growth in earlier chapters. In the growing foetus and baby, growth happens from head to toe (cephalocaudal) and from the trunk outwards to the extremities. Yet in the adolescent, there is a reversal of the proximodistal principle of growth. Instead of growth occurring from the trunk and spreading out to the extremities change first begins in the extremities (hands and feet) and works back towards the trunk. So adolescents' hands and feet grow to their adult size before changes occur to the remainder of their bodies. This is sometimes held accountable for the apparent clumsiness and physical awkwardness seen in young teens who, in theory, have a child's body but adult-sized feet and hands!

Another interesting pattern of growth to note is that this is a period of **asynchronous** growth, which means that changes are rapid and uneven. This is in contrast to the period preceding adolescence when the changes were slow and steady.

The Adolescent Brain

The brain undergoes a major growth spurt, particularly in the frontal cortex which is located at the front of the brain.

Changes in Prefrontal Cortex

Earlier in the book we saw that an infant has far more brain cells and synapses (cell connections) than it needs and a process of 'pruning' occurs, with the brain cutting back the superfluous and inefficient cells and connections. Dr Jay Giedd has discovered that a second wave of synapse formation occurs just before puberty in the prefrontal cortex, whose functions include planning, memory and organisation. A maturing prefrontal cortex increases the teen's ability to reason, to control their impulses and in general to make better judgements. During adolescence 'pruning' again occurs, and the brain discards cells that are not stimulated and used. It's similar to the idea of pruning a tree and cutting back the dead or weak branches. The principle of 'use it or lose it' has been argued to be important during adolescent brain development. If a teen takes part in sport or plays a musical instrument the connections that these experiences stimulate will survive and become 'hardwired' in the brain. If, on the other hand, the teen watches television most of the day those are the cells and connections that will be maintained. Further, an immature prefrontal cortex and its associated functions have been implicated in the risk-taking behaviours seen in some adolescents.

> Now that MRI studies have cracked open a window on the developing brain, researchers are looking at how the newly detected physiological changes might account for the adolescent behaviors so familiar to parents: emotional outbursts, reckless risk taking and rule breaking, and the impassioned pursuit of sex, drugs and rock'n'roll. Some experts believe the structural changes seen at adolescence may explain the timing of such major mental illnesses as schizophrenia and bipolar disorder. These diseases typically begin in adolescence and contribute to the high rate of teen suicide. Increasingly, the wild conduct once blamed on 'raging hormones' is being seen as the by-product of two factors: a surfeit of hormones, yes, but also a paucity of the cognitive controls needed for mature behavior. (Giedd 2004)

Indeed it has been suggested that the brain is not fully mature until 25 years of age. As the physical brain develops and matures we would expect to see the cognitive (or thinking) functions increase and mature as well.

Sexual Maturation

Some of the most obvious characteristics associated with adolescence involve the teenager's changing appearance. Pubertal timing refers to the commencement of puberty and its associated physical and sexual changes. The most noticeable changes in girls are the development of breasts and in boys the appearance of facial hair, for instance. Differentiation is made between primary and secondary sexual characteristics. Let's take a closer look:

Primary Sexual Characteristics

These are the changes and developments that directly involve the sex organs.

Girls — menarche. Menarche refers to the start of menstruation or periods in girls. The timing of puberty for Irish girls (menarche) is on average 13.5 years with a range of 10.9 to 16.1 years of age (cited in Malina *et al.* 2004:314). The average age for menarche in the US is 12.8 years.

Boys — spermarche. This term was coined to mark the sexual maturation of boys: when males become capable of reproduction. 'Spermarche' refers to the first time a boy ejaculates, but as this is a private event it is difficult to surmise an age for this event or to gather much research.

Maturation of reproductive organs:
- in males this includes the growth of the penis and changes to the scrotum and testes
- in females it includes the growth of the vagina, ovaries and uterus.

Secondary Sexual Characteristics

These are visible evidence of changes occurring in the adolescent, e.g.:
- girls — development of breasts
- boys — facial hair and voice change
- both — underarm hair.

Timing of Puberty

While puberty encompasses changes in both males and females we're going to examine the onset of menarche to examine factors involved in the timing of puberty.

Factors in the timing of puberty include heredity and nutrition, and societal influences have also been implicated in the trend of earlier menarche.

Over the last century it has been observed that girls are beginning menarche at increasingly younger ages in the developed world. The age of menarche stabilised in the 1960s at 12.8 years of age in the US and 13.2 in the UK, and in Ireland the mean is 13.5 years of age. Factors implicated in the timing of menarche include weight, exercise and nutrition. Let's take nutrition, for example: if the body experiences a lack of food it may not have the 'fuel' it needs to sustain the immense changes that occur in adolescence and in particular with menarche. Since this is the commencement of reproductive ability in females it makes biological sense, if resources are scarce, for the body to delay possible pregnancy, which would put a further strain on the body's resources.

Garcia-Moro and Hernandez (1990) found in Spanish females in the 1920s an average age of 13.91 years for the start of menarche; this stabilised to 12.92 in the 1950s and has remained at approximately this age since. This is representative of the levelling out of the mean age of menarche that has been witnessed in the West. One would imagine that there has to be a natural cut-off point at which the body cannot begin sexual changes any earlier. Yet we need to remember that not every girl in Ireland begins her period at 13.5 years of age, that a range exists and that there are early and late maturations.

Absent Fathers and Early Pubertal Timing in Girls

Research in the field of pubertal timing has found that several different factors are involved in the start of puberty. Physical influences include weight, nutrition and exercise. For example, high levels of exercise in female dancers have been found to delay pubertal maturation. While physical factors may make intuitive sense to us, some other factors that affect the timing of puberty appear more puzzling.

Ellis and Garber (2000) relates perhaps one of the most fascinating pieces of research to emerge regarding pubertal timing: a relationship between early puberty in girls and the absence of their fathers from their lives.

Ellis and Garber cite Belsky et al. (1991), who found that family environment, including absent *biological* father and maternal depression, related to early pubertal timing in girls. I stress biological father because they found that girls who had stepfathers present in their lives were also more likely to begin puberty early.

Belsky et al. proposed an 'evolutionary' explanation for early pubertal timing. Evolutionary psychology attempts to explain human behaviours by claiming the motivation for these behaviours is found in how we evolved or adapted to our environments. The main focus of evolution is the reproduction of the organism. So the aim of all human behaviour is seen in terms of survival and reproduction. With this in mind the evolutionary explanation of early pubertal timing rests on the notion that a girl whose biological father is absent is more 'vulnerable', has fewer resources and protection; so it makes evolutionary sense for this girl to be able to reproduce earlier than females whose biological fathers are present and caring for them.

Effect of Pubertal Timing

What are the effects of early and late pubertal maturational in females and males? This has been a popular source of research and findings suggest there are different outcomes. Brooks-Gunn *et al.* (1985) suggested that adolescents who are 'off-time' (early or late maturation) with their peers experience more stress and are therefore more susceptible to adjustment problems in adolescence. A second explanation offered is that early developers, especially females, face more social pressure. If they mature earlier they will face the attention of male sexual interest earlier; also, the physical development of their body does not match that of their brain and cognitive development, so they have fewer 'thinking' skills, so to speak, to handle the increased attention that their early maturation attracts. As they look older there can be a tendency for them to be viewed as more socially and cognitively developed than they actually are. It is also suggested that in developing earlier these adolescents miss the opportunity to complete the normal development tasks of middle childhood as they fast forward past this stage.

Early Maturation

In boys this appears to be associated with more positive outcomes, particularly in terms of social development: they enjoy more prestige.

Girls, in contrast to boys, face more problems including an increased likelihood of negative moods and behaviours. Brooks-Gunn *et al.* (1985) reported negative body image.

Higher vulnerability of engaging in risk-taking behaviours (including smoking, drinking and sexual activity) at an earlier age were reported by Magnusson *et al.* (1985).

Late Maturation

Boys showed lower achievement (Dubas *et al.* 1991), lower self-esteem and happiness (Crockett & Petersen 1987).

Higher achievement among girls and lower achievement among boys was evident in research conducted by Dubas *et al.* (1991).

Health Behaviour in School-aged Children (HBSC)

As mentioned in the previous chapter, Health Behaviour in School-aged Children (HBSC) is part of an international research initiative examining health-related behaviours in children and adolescents. According to HBSC Ireland the population of young people (up to 19 years) in the Republic of Ireland is 1,140,616, representing 29 per cent of the population, which is higher than the European average.

The researchers carry out the surveys every four years in an effort to identify trends and gather information about the well-being of Irish children. This research will inform those involved in policy-making, enabling them to put in practice strategies to improve the well-being of children in this country. It also enables us to gauge our levels of general well-being in comparison to the other European countries participating in the HBSC research initiative.

The principal researcher/investigator is the Centre for Health Promotion Studies, University College Galway. In 2002 HBSC Ireland surveyed 8,424 Irish children (age categories 10–11, 12–14 and 15–16) from randomly selected schools throughout the country. They repeated the survey in 2006. Some of their findings will be examined in this book, but more information on HBSC Ireland and their findings can be found on the following websites: www.nuigalway.ie/hbsc/; www.hbsc.org/.

Some Key Findings (HBSC Ireland 2006)

Smoking

Smoking is a leading cause of premature illness and death in developed countries. HBSC Ireland found that 19 per cent of children report that they currently smoke. (Current smoking refers to children who report that they smoke at least monthly.) Those who smoke report feeling less healthy and less happy about their lives compared to non-smokers.

Drunkenness

A greater number of young people are starting to drink at a younger age, by 18 years a higher percentage are regular drinkers, and many are abusing alcohol. HBSC Ireland has found that 31 per cent of children have had so much alcohol that they were really drunk. Children who have been drunk are less likely to have excellent health and to be happy with their lives.

Dieting

HBSC Ireland has found that 13 per cent of children report that they are currently on a diet. (Dieting refers to children who report they are on a diet or doing something to lose weight at present.) This behaviour is of concern among older girls in particular; 24 per cent of 15–17-year-old girls reported that they are currently dieting. The researchers found that children who report being on a diet also reported feeling less likely to find it easy to talk to their parents, less likely to have excellent health or feel happy.

Exercise

Physical activity is associated with good social, emotional and physical health and increasing levels of physical activity, both moderate and vigorous, are recommended. HBSC Ireland has found that 47 per cent of children report that they exercise four or more times a week. (Exercise refers to exercising, outside school hours, enough to get out of breath or sweat.) Children who exercise four or more times a week are less likely to feel pressured by school work, and are more likely to report excellent health and feeling happy.

Food Poverty

Food poverty means being unable to obtain an adequate and nutritious diet. Food poverty tends to affect those from socially disadvantaged groups. HBSC Ireland has found that 16 per cent of Irish children report that they go to school or to bed hungry because there is not enough food at home. Children who report going hungry are less likely to report excellent health and feeling happy while they are more likely to report frequent physical and emotional symptoms.

Health Perceptions

Self-reported measures of health capture an emotional aspect of adolescent well-being. HBSC Ireland has found that 28 per cent of Irish children report excellent health. Children who report having excellent health are more likely to be from a higher social class.

Lifelong Involvement in Sport and Physical Activity (LISPA)

Current rates of participation show that young people's participation peaks around 14 to 15 years of age, with declining rates of activity among adolescents, especially girls. In the chapter on middle age, the LISPA paper and the issue of lifelong participation in sport are dealt with in greater depth.

COGNITIVE DEVELOPMENT

With the increasing activity seen in the brain, one would expect an increase in the cognitive abilities of the adolescent. This growth spurt brings many changes in the way young adults begin to think, including: a better ability to handle more information; and an improved ability to devise mental strategies to help remember and organise information. The growth of the frontal lobe results in greater ability in higher reasoning and thinking. According to Kolb, the growth in the frontal lobe corresponds with Piaget's formal operational stage of thinking, the main features of which are abstract, inductive and deductive thought.

Piaget's Formal Operational Stage (12 Onwards)

The main characteristic of this stage is the growing ability to think in abstract terms, and to use deductive logic. What does this mean?

First let's deal with abstract thinking. Up to this stage the child can only think about things that are 'real' or concrete. According to Piaget they are unable to reason about make-believe problems or situations. Concepts such as justice, ethics or love are abstract terms — you can't touch them or see them. As the adolescent develops abstract thinking they also begin to think more logically; they are able to consider hypotheses (educated guesses or explanations) and test them. (Think of algebra and theorems in school!) Shaffer (1999) reports that 12-year-old children, when asked where they'd put a 'third eye', provide more creative justifications than younger children at the concrete level. One answer at this higher level was a child who would put it on their hand so they could use it to see around corners. Being able to think about their own thoughts reflects their growing cognitive abilities. The ability to think abstractly gives us great freedom in our thinking as we are no longer confined to the world of 'real' objects, concepts or ideas.

Piaget's abstract thinking relates to other areas:

- *neurological* — the increased growth in the prefrontal cortex is responsible for higher reasoning and thought
- *moral* — Kohlberg and Piaget believed that the ability to reason morally at higher and more complex levels came in early teens
- *self* — teens can now think about who they are, their different selves and their possible and future selves, which involves the use of abstract thought.

Do your own research
Ask children and teens of varying ages where they would put a third eye and why. Note their answers and justifications and see if age differences exist.

Check! Learning Goals
You should be able to:

- summarise, in your own words, what abstract thinking means
- explain why the acquisition of abstract thinking is important
- describe how increased cognitive abilities tie in with other areas of development.

Adolescent Egocentrism Theory

Have you observed (or been!) an adolescent girl refusing to leave the house because her make-up wasn't right or because she had a spot? Do you recall the howls that 'everyone' would be looking at her and would see the state of her face? Sound familiar? If it doesn't, count your blessings, but if it rings true David Elkind has an explanation for this behaviour: the **imaginary audience**, which is the belief that others are as concerned about the teen's thoughts and behaviour, and as interested in them, as they are in themselves. It is quite literally the idea that they have an audience watching them. Now, in my thirties, I care far less what others think of me than I did when I was younger. Yet adolescents do care, very much so; and they assume that everyone else is as interested in them as they are. This is why a pimple on the nose turns into a major crisis — the teen believes everyone will notice, everyone will be looking at them. Elkind refers to the increased self-consciousness seen in adolescents as **adolescent egocentrism**. It's hard not to feel sorry for the hapless adolescent going about their daily business in the belief that everyone is watching them.

While the imaginary audience forms one part of the social thinking that characterises adolescent egocentrism, another aspect of it is called the **personal fable**. This is the adolescent's belief that their experiences are unique, and it feeds into a belief that they are not subject to the rules that govern the rest of the world. This contributes to a sense of invincibility. This belief has been implicated in some of the risk-taking behaviours seen in adolescence. Thus a teen might take drugs with the belief that they won't become addicted, that it won't happen to them. While the belief in the imaginary audience is strongest in early teenage years, it declines as the individual moves into later adolescence. The personal fable, on the other hand, can persist into early adulthood.

Check! Learning Goals

You should be able to:

- describe what is meant by 'adolescent egocentrism'
- summarise in your own words what is meant by 'personal fable' and 'imaginary audience', and how they influence/manifest in adolescent behaviour.

SOCIO-EMOTIONAL DEVELOPMENT

David Elkind describes how adolescent behaviours can be explained by their unique style of thinking. Elkind proposes that the 'personal fable' is responsible for the teen's belief in their own invincibility and this leads to risk-taking behaviours. Erikson, who was a socio-emotional theorist, suggests a different explanation for such behaviour. In Chapter 2 we saw that each of Erikson's stages can have either a negative or positive outcome. The negative outcome for the teen, according to Erikson, is 'role confusion' and it is this that is responsible for risky behaviours.

Erikson's Fifth Stage: Identity vs. Role Confusion

As teens make the transition from childhood to adulthood they face questions of 'who' they are and their future role in life. Erikson's fifth stage reflects this — the major developmental 'crisis' to be resolved is that of 'identity'. Adolescence, at least in the West, is synonymous with angst about 'Who am I?' and 'What is my role in life?' and those who successfully navigate this period develop their identity. Erikson referred to the 'psychological moratorium' of the gap between childhood and adulthood, in which the teen explores different roles. By the end of this stage the adolescent will have successfully emerged with a sense of who they are. However, some individuals will be unable to resolve the challenge of emerging identity. According to Erikson they suffer from 'role confusion', resulting either in isolation or in a willingness to take on the identity of others.

Identity is influenced by how the adolescent sees him/herself and is also based on their relationships with others and their perception of how others see them. The concept of identity and self-concept is examined in more depth later in this chapter. For Erikson, the development of self and identity is a major developmental milestone for the teen to accomplish. Those who do not resolve this stage face 'role confusion' which, as the name suggests, leaves the person unsure of who they are and where their lives are going. Role confusion has been linked to risk-taking behaviours.

Check! Learning Goals

You should be able to:

• explain the negative outcome of Erikson's fifth stage and describe its involvement in risk-taking behaviours.
• define a 'psychological moratorium'.

As described by Erikson, identity is the major developmental 'crisis' to be resolved in the transition from childhood to adulthood. Other theories have been put forward to explain how identity emerges. We're going to examine identity and the related concepts of self-concept and self-esteem. First, let's look at another approach to understanding how adolescents characterise self-concept.

Definition of Identity

Identity is the stable, consistent and reliable sense of who one is and what one stands for in the world. It explains one's meaning to oneself and one's meaning to others; it provides a match between what one regards as central to oneself and how one is viewed by significant others in one's life.

Identity is not an unitary concept; it can have many aspects:

- religious
- sexual
- career
- cultural/ethnic.

Who I am in the classroom can be quite different to who I am at home with my son or out with friends: I am Irish, a woman, mother, teacher, friend, daughter and sister.

How do you see yourself? List some of your roles and identities

Marcia's Theories of Identity

Building on the work of Erikson, Marcia (1980) suggested that identity involves the adoption of ideals and values, sexual orientation and work possibilities. Having considered the possibilities, the young person, at the end of adolescence, must make commitments about what to become and what to believe. Marcia used an interview technique to assess identity 'status'. Unlike Erikson, whose theory is a stage-based one, Marcia formulated the idea of status, which allowed for a more fluid conception of identity formation.

Identity statuses are:

- moratorium (in crisis, no commitment)
- achievement (have had crisis and commitment)
- foreclosure (no crisis and commitment)
- diffusion (might have had crisis, no commitment).

Figure 8.1: Marcia's identity statuses

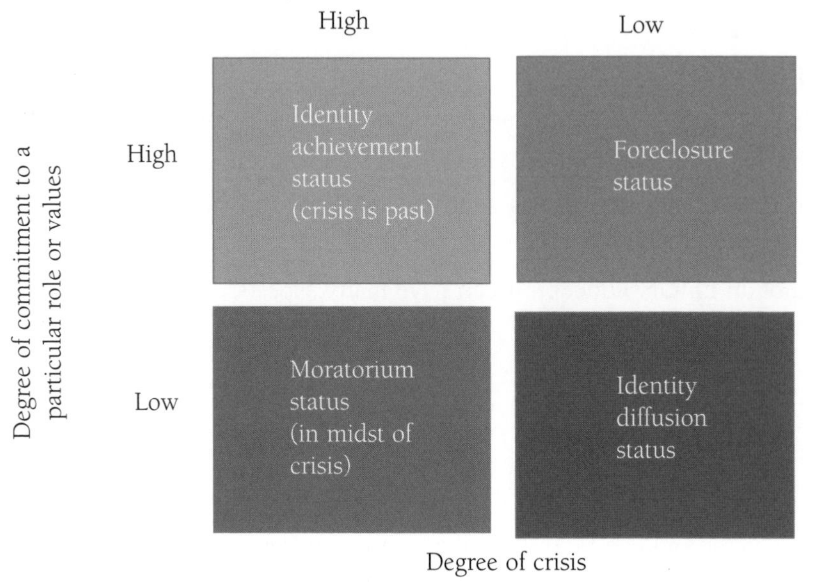

Moratorium

This is marked by intense crisis as the teen searches out the possibilities, considering different roles in their search for identity. No commitment has been made yet.

Identity Diffusion

The main characteristic of this stage is that the teen is not in 'crisis', actively considering the possibilities, further they are avoiding making a commitment. The stereotypical difficult teen watching television and refusing to talk might be said to be in this state.

Identity Foreclosure

This teen might seem like the perfect adolescent who dutifully doesn't question or rebel. This teen has made a commitment to an identity without going through the crisis and exploring alternatives. Instead they have accepted a parental or cultural defined commitment. Erikson would believe that the teen had not tackled the challenges of this period and was relying on identification with others rather than carving out their own identity.

Identity Achievement

The adolescent has come through the 'crisis' and a commitment has been made to ideological, occupational and other goals. They have established a sense of self, of their identity.

Check! Learning Goals

You should be able to:

- list the four identity statuses
- explain the difference between 'crisis' and 'commitment'.

Self-Concept and Self-Esteem

Another term for identity is self-concept. This can refer to all aspects of the self. We compare ourselves to others and how we evaluate or judge aspects of ourselves is called self-esteem. Harter (1990) defined self-esteem as 'how much a person likes, accepts, and respects himself [sic] overall as a person' (p. 255). During adolescence self-esteem can be fragile: Elkind's imaginary audience indicates that teens have the potential to judge themselves mercilessly. Why have researchers who are interested in identity development paid so much attention to adolescence?

In the section on physical development we examined the changes that occur with puberty, and these changes can alter the teen's self-conception. Piaget maintained that adolescents are now able to think at far higher levels, and this increased ability allows them to appreciate fully just how significant the changes are. The acquisition of abstract thinking allows them to imagine themselves in the future or in different roles. This feeds into greater self-consciousness. We shall see later in this chapter that low self-esteem has been implicated in eating disorders.

The area of self-esteem has been extensively researched, including the relationship between low self-esteem and low life satisfaction, loneliness, anxiety, resentment, irritability and depression (Rosenberg 1985). High self-esteem has been found among teens who saw themselves as close to their parents (Blyth & Traeger 1988).

An excellent book written by Orla McHugh, *Celtic Cubs: Inside the Mind of the Irish Teenager*, provides an interesting analysis of Irish society and its influence on teen development. The book includes a helpful section on parenting.

In Focus: Eating Disorders

As we saw, the HBSC Ireland team looked at dieting, and this brings us to eating disorders, an issue often associated with adolescence. Eating disorders, while they manifest themselves in the physical domain (weight loss and disturbed eating patterns), are seen as a psychological disorder. Examining them offers us a bridge between the physical and psychological domains of functioning.

Anorexia Nervosa

Anorexia is a Greek word meaning loss of appetite. Anorexia nervosa is an eating disorder that causes self-starvation in millions of people each year. This disorder allows adolescents to gain control by limiting food intake. Anorexics often obsess about thinness, need attention, lack individuality, and deny sexuality. It affects fifteen times more females than males. It usually begins during adolescence or in early adulthood, but hardly ever occurs in women past the age of 25. The disorder affects one teenager in every 200 among adolescents aged between 16 and 18.

The **criteria** for the diagnosis of anorexia nervosa given by the most recent diagnostic system include:

- intense fear of becoming obese, which does not diminish with the progression of weight loss
- disturbance of body image, feeling 'fat' even when emaciated
- refusal to maintain body weight over a minimal weight for age and height
- weight loss of 25 per cent of original body weight or 25 per cent below the expected weight based on standard growth charts
- no known physical illness that would account for the weight loss (Field and Domangue 1987:31).

Bulimia

Bulimia is derived from the Greek words meaning 'ox' and 'hunger': the sufferer eats like a hungry ox. A bulimic eats in unrestrained sprees. One major characteristic of bulimia is binge eating, in which the bulimic can consume between 10,000 and 20,000 calories. These binge eating episodes are usually followed by episodes of purging, accomplished by any of the following methods: vomiting, laxatives, diuretics, enemas, compulsive exercising, weight-reducing drugs, and intermittent periods of strict dieting. Field and Domangue define it as a syndrome in which gorging on food alternates with purging by forced vomiting, fasting or laxatives (1987:32).

Bulimia can be associated with anorexia nervosa. In the past it was considered to be a part of the same disorder. Most anorexics develop bulimia in the course of their illness. Bulimia affects three to seven per cent of women aged between 15 and 35. The effects on people who suffer from bulimia include guilt, depression and self-disgust. Bulimics know that their eating habits are unhealthy but fear not being able to stop eating voluntarily.

Both anorexia nervosa and bulimia involve an obsession with weight and body image, and both are usually found in white, middle- and upper-class females. These women are usually perfectionists, high achievers, often academically or vocationally successful, and have a great need to please others.

The mortality rate in eating disorders is higher than for any other psychiatric disorder, and studies suggest that the rate has been increasing over the last 20 years.

Mental Attributes of Teenagers with Bulimia and Anorexia Nervosa

It is important to understand that adolescents suffering from eating disorders, both male and female, may not appear to be underweight. Weight is only a physical sign of an eating disorder, but the person is likely to be suffering from a deeper emotional conflict that needs to be resolved. Eating disorders are addictions to a behaviour and the obsession with food is a symptom of deeper problems such as low self-esteem, depression, poor self-image, and self-hate. Hilde Bruch, an eminent psychiatrist, describes the relentless pursuit of thinness as an effort to mask underlying problems (Bruch *et al.* 1988.) The following are some of the mental characteristics of adolescent males and females suffering from eating disorders.

Perfectionism. Many adolescents suffering from eating disorders are perfectionists and high achievers. They strive to reach perfection in every aspect of their lives. They are eager to please and in the process lose their true selves. When they feel they have failed to reach perfection they often unrealistically blame themselves for their failure and attempt to punish themselves. Punishment often occurs in the form of starvation for anorexics and purging for bulimics.

Low self-esteem. Feelings of inadequacy are common among teens suffering from eating disorders. They have a poor self-image and perception of themselves. They irrationally believe they are fat, regardless of how thin they become. They experience a sense of inner emptiness, uncertainty and helplessness, and a lack of self-confidence and self-trust (Bruch *et al.* 1988). Often they are afraid of being judged by others or thought of as being stupid. They feel confident if they are losing weight but suffer from feelings of worthlessness and guilt if they are not (Pipher 1994). It is also common for them to believe they do not deserve good things or to be happy.

Depression. Mood swings, feelings of hopelessness, anxiety, isolation and loneliness are feelings common to sufferers of eating disorders. Bulimics experience a loss of control that leads to depression, whereas anorexics experience depression as a result of gaining weight (Schlundt 1990).

Obsession. These adolescents deal with an intense obsession and preoccupation with food, calories, fat grams and weight. Weight becomes their most important and self-defining attribute. Eating disorders are considered addictions in which starvation, bingeing and purging are the addictive behaviors and food is the narcotic (Pipher 1994).

Guilt. Adolescents with eating disorders often feel guilty because they do not think they have met the expectations of others. They are striving for the perfect body and for a sense of control but in the process they start to feel guilty about their habit (Pipher 1994). Lying becomes essential.

Treatment

According to the Irish College of Psychiatrists' report, 'A better future now', treatment principles for eating disorders are as follows:

- The treatment of anorexic patients must be based on a comprehensive and detailed assessment of their mental and physical status.
- A decision is then made as to whether treatment in an in-patient or outpatient setting is more appropriate.
- The increase in clinical cases has affected admission rates not only to psychiatric units but also to paediatric medical wards.
- It is clear that a severely emaciated patient will require intensive medical treatment; it can be extremely difficult to provide the necessary psychotherapeutic milieu in such an environment.
- In-patient treatment in a specialist unit offers the attraction of a wealth of accumulated experience in treating these disorders; however, patients with eating disorders can be effectively treated in a unit that treats a range of psychiatric disorders.
- Working with the parents is an integral part of any intervention programme.
- Outpatient programmes are increasingly considered as a treatment option; these involve specialist teams, who may work jointly with community teams.

(Irish College of Psychiatrists 2005:78)

In Focus: Suicide

Definition: the wilful taking of one's own life.

Some Facts

- The World Health Organisation (WHO) estimates that 500,000 people worldwide commit suicide annually, about 1.4 every minute.
- In the USA, suicide rates among 15- and 24-year-olds have trebled since 1960.
- Women make about three times as many suicide attempts as men, but men are three times more likely to actually kill themselves. These differences may be due to:
  - a higher incidence of depression in women
  - men's choice of more lethal methods, such as shooting themselves or jumping off buildings.
- The rate for both genders is higher among those who have been divorced or widowed.
- Women who commit suicide have a relatively greater tendency to be motivated by failures in love relationships, whereas men have a greater tendency to be motivated by failure in their occupations.
- A history of sexual or physical abuse significantly increases the likelihood of later suicide attempts (Garnefski & Arends 1998).
- Depression is one of the strongest predictors of suicide; approximately 15 per cent of clinically depressed individuals will kill themselves.

Motives

There appear to be two fundamental motivations for suicide attempts:

1. the desire to end one's life
2. the desire to manipulate and coerce other people into doing what the suicidal person wants.

In one study, 56 per cent of suicide attempts were classified as having been motivated by the desire to die (Beck 1976).

Parasuicide is an attempt that does not end in death, often seen as a cry for help or an attempt to force other people to meet one's needs.

An Irish Perspective

Prevalence:

- Ireland ranks fifth highest in the EU for rates of youth suicide (14–24-year-olds), and although the precise rates of mental health problems for young people can be difficult to calculate, it is estimated that the prevalence of mental health difficulties for children and young people can be up to 20 per cent at any one time.

- For every woman who commits suicide approximately four to five men will.

- Figures for 2003 showed that there were 12.5 suicide deaths per 100,000 of the population in Ireland, compared with 8.5 such deaths in the North. Comparative figures for the UK in the same year were: England 9.9 deaths; Wales 12 deaths; and Scotland 15.7 deaths per 100,000 — the highest rate in these islands.

- Mr Derek Chambers, research and resource officer of Ireland's National Suicide Review Group, pointed out that, among people in their 20s or early 30s, the suicide rate was higher, at 15.7 deaths per 100,000. He said that while almost one in four suicide victims was reported by a doctor or psychiatrist to have a history of alcohol abuse, a US expert on suicide prevention claims that 90 per cent of people who took their own lives had 'diagnosable mental or substance-abuse disorders, or both'. (*Irish Times*, 26 August 2005)

The following extract is from a study entitled *The Male Perspective: Young Men's Outlook on Life* by the National Suicide Research Foundation.

From a sociological perspective:

A sense of community (social integration) and shared values (social regulation) can influence the behaviour and actions of individuals. In this context, the social changes that have occurred in Ireland in recent years merit investigation in terms of their impact on men's sense of personal worth and belonging in our modern society. Previous research has speculated that sociocultural changes in Western societies in recent years have adversely affected men more than women and that a gender difference has emerged in terms of how 'the self is seen or construed'.

Furthermore, increased individualism may be contributing to a greater sense of isolation for young men, as women tend to remain more socially connected and to view the self as interdependent with others whereas men are more likely to view the self as separate.

Psychologically:

The individual's perceived sense of control is also important in determining how problems are dealt with and challenges in life are met. By clarifying young men's existing sense of control, realistic and meaningful health promotion strategies may be identified.

Anomy:

A concept that has been applied to the understanding of changes in suicide rates is the concept of anomy. Anomy describes the unbalancing of social forces that affect individual action. It implies an upsetting of the balance or normality in a previously accepted way of life. It is based on the notion that society usually exercises control over individual behaviour and desire through social rules and norms and when these rules and norms break down individual behaviour is no longer regulated by society. At an individual level it can be described as a personal feeling of not being part of, or responsible to, society. Underlying the notion of anomy is the belief that human desire is basically infinite, human beings will always want more unless society controls desires through the existence of everyday shared values and institutionalised rules or laws. Without a sense of accepted social values individual behaviour or desire may not be controlled or regulated and the level of so called deviant or unacceptable behaviour, including suicide, increases.

At an individual level:

Changes in personal circumstances can lead to uncertainty or can upset the normal way of life. For example, by winning a large sum of money an individual may be forced to question values and desires that were previously taken for granted as new opportunities present themselves. Similarly, negative events such as job loss or divorce can upset the balance or equilibrium that previously governed an individual's way of life.

Anomy is also related to the values and expectations in a society and the means to achieve these expectations. When there is a discrepancy between expectations and means to achieve them then the level of anomy increases. For example, in Irish society home ownership is something that is valued and is associated with independence and passage into adulthood. However, the means to achieve home ownership are not readily available to the majority of young people attempting to make the transition from adolescence to adulthood. The resulting situation may be that an individual is achieving societal expectations on some levels, e.g. in terms of career, but not on other levels such as home ownership.

While anomy contributes to our overall understanding of suicide as a social problem, as a symptom of social transition, it may be difficult to apply to the understanding of individual deaths by suicide.

(Begley *et al.* 2004)

Sources of Information
* National Suicide Research Foundation: www.nsrf.org
* Department of Health: www.doh.ie
* Health Research Board (includes good links to information on drugs): www.hrb.ie

Useful Contacts
Aware run a helpline counselling service for people with depression and their families, 24 hours a day, seven days a week.
Helpline: 1890 303302
72 Lower Leeson Street Dublin 2
Tel: 01 661 7211, email: *aware@iol.ie*, website: www.aware.ie.

The Samaritans provide confidential 24-hour emotional support for people experiencing feelings of distress or despair, including those that may lead to suicide.
Helpline: 1850 609090
112 Marlborough Street, Dublin 1
Tel: 01 872 7700.

CHAPTER SUMMARY

Adolescence encompasses behavioural, social and emotional aspects, which are influenced by physical and social/cultural factors, during the age range ten–20 years of age.

Puberty refers to the physical and sexual changes that occur during this period.

Physical Development
The growth spurt occurs:
* in girls — at approximately ten years of age

- in boys — at approximately 12 years of age.

Motor skills: girls tend to be better co-ordinated in their movements, while boys may have greater speed and strength.

Growth Patterns

There is a reversal of the proximodistal principle of growth: change begins in the extremities (hands and feet) and works back towards the trunk.

Adolescence is a period of **asynchronous** growth — changes are rapid and uneven.

The Adolescent Brain

The brain undergoes a major growth spurt, particularly in the prefrontal cortex, which is located at the front of the brain.

Changes in prefrontal cortex: a maturing prefrontal cortex increases the teen's ability to reason, to control their impulses and in general to make better judgements. It is suggested that the brain is not fully matured until 25 years of age.

Sexual Maturation

Pubertal timing refers to the beginning of puberty and its associated physical and sexual changes.

Primary sexual characteristics — changes and development that directly involve the sex organs.

Menarche refers to the start of menstruation or periods in girls. The timing of puberty (menarche) for Irish girls is 13.5 years with a range of 10.9 to 16.1 years of age.

Spermarche. This term was coined to mark the sexual maturation of boys, when males become capable of reproduction.

Maturation of reproductive organs:
- in males — growth of the penis and changes to the scrotum and testes
- in females — growth of the vagina, ovaries and uterus.

Secondary sexual characteristics are visible evidence of changes occurring in the adolescent:
- girls — development of breasts
- boys — facial hair and voice change
- both — underarm hair.

Differences in Timing of Puberty

Factors in the timing of puberty include heredity and nutrition.

Girls are beginning menarche at increasingly younger ages in the developed world. Factors implicated in the timing of menarche include weight, exercise and nutrition.

Absent fathers have also been implicated as a factor in early pubertal timing. An 'evolutionary' explanation of early pubertal timing rests on the notion that a girl whose biological father is absent is more 'vulnerable', has fewer resources and less protection, so it makes evolutionary sense for this girl to be able to reproduce earlier than females whose biological fathers are present and caring for them.

Effect of Pubertal Timing

Early maturation:

- boys — associated with more positive outcomes, particularly in terms of social development; they enjoy prestige
- girls — more problems, including increased likelihood of negative moods and behaviours, negative body image and of engaging in risk-taking behaviours.

Late Maturation:

- boys — lower achievement, self-esteem and happiness
- girls — higher achievement.

Health Behaviour in School-aged Children (HBSC)

Some of the key findings from the HBSC Ireland 2006:

- **Smoking** — 19 per cent of children report that they currently smoke.
- **Drunkenness** — 31 per cent of children have had so much alcohol that they were really drunk.
- **Dieting** — 13 per cent of children report that they are currently on a diet; 24 per cent of 15–17-year-old girls reported that they are currently dieting.
- **Exercise** — 47 per cent of children report that they exercise four or more times a week.
- **Food poverty** (when one cannot obtain an adequate and nutritious diet) — 16 per cent of Irish children report that they go to school or to bed hungry because there is not enough food at home.
- **Health perceptions** — 28 per cent of Irish children report excellent health.

Cognitive Development

Piaget's Formal Operational Stage (12 Onwards)

Main characteristics:

- abstract thinking, no longer confined to 'real' objects, concepts or ideas
- deductive logic: adolescents begins to think more logically and are able to consider and test hypotheses (educated guesses or explanations).

Elkind's Theory of Adolescent Egocentrism

The increased self-consciousness seen in adolescents is referred to as **adolescent egocentrism** by Elkind. **Imaginary audience** is the adolescent belief that others are as concerned about their thoughts and behaviour as they are. It is quite literally the idea that they have an audience watching them.

The **personal fable** is the adolescent's belief that their experiences are unique and that they are not subject to the rules that govern the rest of the world. This contributes to a sense of invincibility.

Socio-emotional Development

Erikson's Fifth Stage: Identity vs. Role Confusion

As teens make the transition from childhood to adulthood they face questions of 'who' they are and their future role in life.

Adolescents who are unable to resolve the challenge of emerging identity face **role confusion**, resulting in either isolation or a willingness to take on the identity of others.

Identity is 'the stable, consistent, and reliable sense of who one is and what one stands for in the world'.

Marcia's Identity Statuses and Development

Marcia suggested that identity involves the adoption of ideals and values, sexual orientation and work possibilities. He formulated the idea of status, which allowed for a more fluid conception of identity formation.

Identity statuses:

- moratorium (in crisis, no commitment)
- achievement (have had crisis and commitment)
- foreclosure (no crisis and commitment)
- diffusion (might have had crisis, no commitment).

Self-concept and Self-esteem

Self-concept refers to all aspects of the self.

Self-esteem is defined by 'how much a person likes, accepts, and respects himself overall as a person'.

There is a relationship between low self-esteem and low life satisfaction, loneliness, anxiety, resentment, irritability and depression.

High self-esteem was found amongst teens who saw themselves as close to their parents.

CHAPTER NINE

ADULTHOOD

The period of adulthood has not received the same degree of attention or study as childhood and adolescence. Things are beginning to change, however, with more research being devoted to this once ignored area.

CHAPTER OUTLINE

The chapter is divided into three sections:

- Early Adulthood (20–40)
- Middle Adulthood (40–65)
- Late Adulthood (65 onwards).

Each section looks at the physical, cognitive and socio-emotional domains as with previous chapters. Before discussing the age-specific categories a general introduction to ageing and related concepts is necessary. We will look at theories more closely as we move from early to later adulthood.

First we shall examine some of the concepts relating to ageing and also general policy and research on the health of the Irish nation.

AGEING

There are many theories of ageing, but first the distinction needs to be made between primary and secondary ageing.

Primary ageing refers to the gradual age-related changes over which we have no control, reflecting the biological aspects of ageing that affect everyone. Changes include greying hair and deterioration in hearing and vision.

Secondary ageing is more individual: it is influenced by lifestyle and environmental factors such as lack of exercise, smoking, alcohol consumption and obesity. Disease also features as a factor in secondary ageing.

Theories of Ageing

Damage-based theories suggest that ageing results from a continuous process of damage which accumulates throughout the entire lifespan. This approach argues that ageing is predominantly a result of interactions with the environment.

Programmed theories of ageing suggest that ageing is not a result of random processes but rather that ageing is predetermined and occurs on a fixed schedule.

Signs of Senescence

Senescence is the deterioration of bodily functions that accompanies ageing in a living organism. Some signs of ageing include:

- an overall decrease in energy
- the tendency to become easily tired
- wrinkles and/or brown spots on the skin and loss of skin elasticity
- greying and thinning of hair
- loss or decrease in vision and hearing
- slower reaction times.

IRISH HEALTH RESEARCH AND POLICY

We are going to look at two areas: first the National Health Promotion Strategy 2000–2005, which outlines the intended areas for policy development and intervention. Then we will examine the Survey of Lifestyle, Attitudes and Nutrition (SLAN), which surveys the adult Irish population every four years to obtain data regarding health and lifestyle behaviours.

The National Health Promotion Strategy 2000–2005

This strategy was launched with the aim of improving the health and well-being of the nation. The purpose of the strategy was to set out a broad policy framework within which action can be carried out at an appropriate level to advance its strategic aims and objectives (see Appendix 1 of the document). The full text can be accessed at www.dohc.ie/publications/pdf/hpstrat.pdf?direct=1

When defining the health promotion strategy it became clear that data were needed to ascertain trends and to identify the health behaviours of Irish people, and SLAN was implemented to meet these needs.

Survey of Lifestyle, Attitudes and Nutrition (SLAN)

In earlier chapters on childhood and adolescence, we examined the Health Behaviour in School-aged Children (HBSC) survey and their Irish findings. A parallel survey is examining the health and lifestyle of Irish adults aged over 18. The SLAN survey examines topics across age categories (18–34, 35–54 and 55 onwards) and gender. Educational status and social class are also categorised, thus giving a fuller picture of the relationship between these variables or factors.

In building a base of knowledge, it is envisaged that these findings will allow for trends to be identified and for the implementation of policy to meet these needs.

Let's take a look at some of the findings.

Findings

Smoking

Currently one in three adults are smokers, and smoking rates among younger women are now comparable with those among men. There is a strong negative social class gradient in smoking prevalence in all age groups of both men and women. About 20 per cent of all older women smoke. Most smokers want to quit but perceive the lack of willpower as the main problem.

Food and Nutrition

Thirty-two per cent of adults reported a Body Mass Index (BMI) classifiable as overweight and ten per cent as obese. (BMI is the ratio of weight to height squared and is used to classify people into the following categories: normal (BMI up to 25), overweight (BMI 25 to 29) and obese (BMI greater than or equal to 30).

Alcohol

There has been a shift in patterns of drinking in that most adults now drink alcohol. Twenty-seven per cent of adult males and 21 per cent of adult females consume more than the recommended weekly limits of sensible alcohol consumption. Twenty-two per cent indicated that they had driven after having consumed two or more alcoholic drinks.

Exercise

Overall 42 per cent of adult respondents engaged in some form of regular physical exercise. Rates declined markedly with age. Nearly one third of those over 55 years took no exercise at all in a typical week. A social class gradient existed in most age groups for both males and females.

For adults, injury varied according to gender and age. Young males were most likely to report a sports-related injury (41 per cent). Those in the 35–54-year age group were most likely to report a work-related injury (37 per cent) and the oldest age group, 55 years or over, reported home- or garden-based injuries (48 per cent).

EARLY ADULTHOOD

An outline of this section:
- Physical development:
  - The brain
  - The ageing process
  - SLAN findings
- Cognitive development:
  - Schaie's theory of cognitive development
  - Postformal thinking
  - Fluid and crystallised intelligence
- Socio-emotional development:
  - Erikson's isolation vs. intimacy
  - Levinson's seasons of man
- Trends in having children.

Physical Development
The Brain

A second major spurt in the brain occurs at approximately 17 years of age and continues until age 21 or 22. Most of the changes occur in the frontal lobe, which is involved in higher reasoning and emotional control.

The Ageing Process

According to the 'wear and tear' approach in understanding ageing, the body simply wears out over time. Hayflick (1980) found that human cells are set to reproduce themselves a fixed or finite number of times, after which they will eventually die.

Changes can occur to the cells: **free radicals**, which we hear about so often, alter and damage cells. Environmental factors such as pollution and smoking encourage the production of free radicals. Antioxidants, such as vitamins A and E, found in fruit and vegetables, can neutralise these. Nonetheless free radical damage accumulates with age.

This highlights the importance of a healthy diet and lifestyle in maintaining good health and longevity. Bill Creasey, coach education officer and coach tutor, further elaborates on the importance of lifelong physical activity and its many benefits in the 'In Focus' section on lifelong participation below.

SLAN Findings

General Health

The majority of respondents, both male and female, reported good health, with a minority (one per cent) reporting 'poor' health in both the 1998 and 2002 surveys. The top five requirements identified by respondents for improving health were:

- less stress
- more willpower
- change in weight
- more money
- less pollution.

Smoking

In the 18–34 age category, the numbers of regular or occasional smokers were:

	1998	**2002**
Males	38%	35%
Females	40%	33%
Overall	39%	34%

There was a decrease in reported smoking in the period from the 1998 survey to the 2002 survey. Yet, in comparison to the other two other age categories, younger people continue to smoke more.

In general, those with lower educational attainment reported smoking more than those who had completed secondary school and/or third-level education. According to the authors this is in keeping with international literature.

Cognitive Development

Schaie's Theory of Cognitive Development

K. Warner Schaie (2000) offers a theory of cognitive development across the lifespan, beginning with 'acquisition' and ending towards the end of life with 'legacy creating'. In early adulthood the stages are:

- **Acquisition** (childhood and adolescence). This involves the acquisition of mental structures, information and skills in order to gain understanding of the world.
- **Achieving** (late teens to early twenties). Young adults use their knowledge to pursue a career, choose a lifestyle, and solve personal dilemmas.

Check! Learning Goals

You should be able to:

- describe the difference between 'acquisition' and 'achieving' and identify their age ranges.

Postformal Thinking

It has been argued that a fifth stage exists after Piaget's formal operational stage (see Chapter 8). Rather than the mere acquisition of mental structures, the individual is now able to apply this new level of thinking to problem solving and reasoning. **Postformal thought** incorporates the ability to recognise that the correct answer to a problem requires reflective thinking and may vary from situation to situation. Adolescents tend to think in a dualistic manner, such as right or wrong; adults recognise that shades of grey exist and that each situation or problem is unique.

Check! Learning Goals

You should be able to:

- explain how postformal thought differs from Piaget's formal operational stage.

Fluid and Crystallised Intelligence

Raymond Cattell and John L. Horn combined to produce a new understanding of the course of intellectual development across the lifespan. Their theory proposed that general intelligence is not a unitary factor but instead a grouping of approximately a hundred abilities working together. These numerous abilities are separated into two strands or abilities (fluid and crystallised) that have quite different developmental pathways across the lifespan.

Fluid intelligence drives the individual's ability to think and act quickly and to solve problems. Fluid intelligence is grounded in physiological efficiency, and is thus relatively independent of education or cultural influences (Horn & Cattell 1967). Many studies have demonstrated that fluid intelligence peaks in early adulthood and then declines, gradually at first and then more rapidly as old age sets in after about 70 years of age.

Crystallised intelligence stems from learning and culture influences, and is reflected in tests of knowledge, general information, use of language (vocabulary) and a wide variety of acquired skills (Horn & Cattell 1967). Personality factors, motivation and educational and cultural opportunity are central to its development. Crystallised abilities continue to improve as individuals age. Crystallised intelligence can be thought of as wisdom that the individual accumulates as they progress through life.

Check! Learning Goals

You should be able to:

* describe fluid and crystallised intelligences
* explain which intelligence improves with age.

Socio-emotional Development
Erikson's Isolation vs. Intimacy (20–30 years)

In this stage Erikson proposes a positive outcome (intimacy) and a negative one (isolation). Intimacy relates to forming an intimate or love relationship and committing to it. Erikson suggested that isolation reflected an individual's hesitation to form a relationship, possibly due to a fear of losing identity or to self-absorption.

What do you think? Have the changes in Irish society over the last 50 years affected people's ability and opportunity to meet and form love relationships?

Levinson's Seasons of Man

Daniel Levinson (1986) devised a theory that extends over the entire lifespan but pays particular attention to the 'nature' of adult development. Central to his theory is the idea of four 'seasonal cycles': pre-adulthood; early adulthood; middle adulthood; and late adulthood. The scheme was initially based on interviews with 40 men. Levinson later interviewed women and found they proceeded through the same eras (life cycles) in a similar manner to men. Levinson did find, though, that women's lives are more closely linked to the family cycle.

Underlying Levinson's theory are some important concepts.

A **life structure** refers to the underlying pattern of an individual's life at a given time and is shaped by their social and physical environment, primarily family and work, although other factors such as religion and economic status are often important.

A **life cycle** relates to the underlying order spanning adult life, consisting of a sequence of eras. The move from one era to another is not smooth; there are cross-era **periods of transition** which last for about five years.

Figure 9.1: Levinson's seasons of man

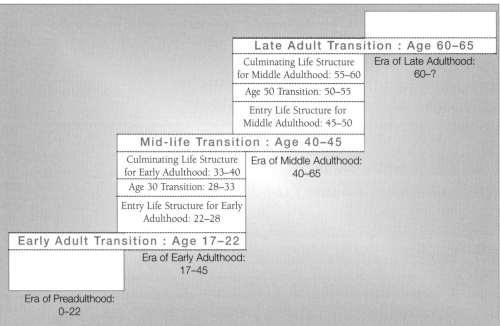

As you can see from the image, according to Levinson (1986:7), there are nine developmental periods. The stages relating to early adulthood are:

- **Early Adult Transition (17–22)** is a developmental bridge between pre-adulthood and early adulthood.

- **Entry Life Structure for Early Adulthood (22–28)** is the time for building and maintaining an initial mode of adult living.
- **Age 30 Transition (28–33)** is an opportunity to reappraise and modify the entry structure and to create the basis for the next life structure.
- **Culminating Life Structure for Early Adulthood (33–40)** is the vehicle for completing this era and realising our youthful ambitions.
- **Midlife Transition (40–45)**, is another of the great cross-era shifts, serving both to terminate early adulthood and to initiate middle adulthood.

As we can see, starting careers and families marks the beginning of early adulthood. Around age 30, an evaluation occurs and people begin to settle down and also work towards career advancement. Another transition begins at age 40 with the realisation that not all the ambitions and goals that were set will be achieved.

Check! Learning Goals
You should be able to:
- outline the difference between a 'life structure' and a 'life cycle'
- list the five stages of early adulthood according to Levinson
- describe at least two of the stages in your own words.

Trends in Having Children
According to figures released by the Central Statistics Office (CSO 2006b), the average age of first-time mothers continues to rise. The average age of first-time mothers in 2006 was 28.7 years, compared to 27.1 years of age for first-time mothers in 1997 and 25.0 years in 1977. In 2006, just under one-third of births were outside marriage.

MIDDLE ADULTHOOD
An outline of this section:
- Physical development:
  - Menopause
- Cognitive development:
  - Schaie's stages of cognitive development
  - Sternberg's triarchic theory of successful intelligence
- Socio-emotional development:

- Erikson's generativity vs. stagnation
- Levinson's middle adulthood
- Happiness
- Parenting: single mothers and psychological well-being.

Physical Development

Menopause

The menopause marks the end of the reproductive capacity of the woman as her periods cease. The age range for the menopause is between 40 and 60 years of age, though it generally occurs when a woman is in her fifties. It is considered the most significant physical change experienced by women in middle adulthood.

Cognitive Development

Schaie's Stages of Cognitive Development

- **Responsibility** (middle adulthood). Individuals use their abilities to solve problems related to their responsibilities for others, for example family members.
- **Executive** (middle adulthood). This stage reflects a concern for the welfare of the broader social system. People deal with complex relationships on multiple levels.

Check! Learning Goals

You should be able to:

- explain, in your own words, the characteristics of Schaie's two stages in middle adulthood.

Sternberg's Triarchic Theory of Successful Intelligence

Robert J. Sternberg (1949–) argues that intelligent behaviour derives from a balance between analytical, creative and practical abilities. Intelligent behaviour is reflected in optimal adaptation to your environment or particular sociocultural context (Sternberg 1988). The three abilities are:

- **analytical** — enable the individual to evaluate, analyse, compare and contrast information
- **creative** — generate invention, discovery and other creative endeavours
- **practical** — tie everything together by allowing individuals to apply what they have learned in the appropriate setting.

Figure 9.2: Age changes in physical functioning (Boyd & Bee 2005)

Body Function	Age at which change begins to be clear or measurable	Nature of Change
Vision	Mid-40s	Lens of eye thickens and loses accommodative power, resulting in poorer near vision and more sensitivity to glare
Hearing	50 to 60	Loss of ability to hear very high and very low tones
Smell	About 40	Decline in ability to detect and discriminate among different smells
Taste	None	No apparent loss in taste discrimination ability
Muscles	About 50	Loss of muscle tissue, particularly in 'fast twitch' fibres used for bursts of strength or speed
Bones	Mid-30s (women)	Loss of calcium in the bones, called osteoporosis; also wear and tear on bone in joints, called osteoarthritis, more marked after about 60
Heart and lungs	35 to 40	Most functions (such as aerobic capacity or cardiac output) do not show age changes at rest, but do show age changes during work or exercise
Nervous system	Probably gradual throughout adulthood	Some loss (but not clear how much) of neurons in the brain; gradual reduction in density of dendrites; gradual decline in total brain volume and weight
Immune system	Adolescence	Loss in size of thymus; reduction in number and maturity of T cells; not clear how much of this change is due to stress and how much is primary ageing
Reproductive system	Mid-30s (women)	Increased reproductive risk and lowered fertility
	Ealy 40s (men)	Gradual decline in viable sperm beginning at about age 40; very gradual decline in testosterone from early adulthood
Cellular elasticity	Gradual	Gradual loss of elasticity in most cells; including skin, muscle, tendon, and blood vessel cells; faster deterioration in cells exposed to sunlight
Height	40	Compression of discs in the spine, with resulting loss of height of 1 to 2 inches by age 80
Weight	Nonlinear	In US studies, weight reaches a maximum in middle adulthood and then gradually declines in old age
Skin	40	Increase in wrinkles, as a result of loss of elasticity; oil-secreting glands become less efficient
Hair	Variable	Hair becomes thinner and may grey

To be successful in life the individual must make the best use of their analytical, creative and practical strengths, while at the same time compensating for weaknesses in any of these areas. For example, a person with highly developed analytical and practical abilities, but with less well-developed creative abilities, might choose to work in a field that values technical expertise but does not require a great deal of imaginative thinking. Thus, a central feature of the triarchic theory of successful intelligence is adaptability — both within the individual and within the individual's sociocultural context (Cianciolo & Sternberg 2004).

Check! Learning Goals

You should be able to:

• describe the three abilities that form Sternberg's theory
• explain why Sternberg calls it a theory of 'successful' intelligence.

Socio-emotional Development

Erikson's Generativity vs. Stagnation

Generativity comes from the word 'to generate' and Erikson envisaged this stage as offering the opportunity to the individual to reach out to others and guide the younger generation. This can be seen particularly in child-rearing but also in more civic-minded activities such as coaching or community involvement. Here the focus lies beyond the self and on others. If this does not occur the individual is faced with feelings of stagnation and places their needs above challenge and sacrifice. They are self-indulgent, displaying little interest in work productivity or involvement with younger people.

Levinson's Middle Adulthood

• **Midlife Transition (40–45)** is another of the great cross-era shifts, serving both to terminate early adulthood and to initiate middle adulthood.
• **Entry Life Structure for Middle Adulthood (45–50)**, like its counterpart above, provides an initial basis for life in a new era.
• **Age 50 Transition (50–55)** offers a mid-era opportunity for modifying and perhaps improving the entry life structure.
• **Culminating Life Structure for Middle Adulthood (55–60)** — the framework in which we conclude this era.

- **Late Adult Transition (55–60)** is a boundary period between middle and late adulthood, separating and linking the two eras.

Check! Learning Goals

You should be able to:

- compare and contrast Erikson's and Levinson's stages
- list Levinson's stages and explain them in your own words.

Happiness

Ed Diener (1946–) is a professor of psychology at Illinois University and is responsible for some of the most innovative research in the area of happiness. Professor Diener is part of a movement in psychology called 'positive psychology', in which interest lies not in the traditional deficit model approach to understanding psychology but instead in examining factors that are involved in positive behaviours, thoughts and feelings. Some of the findings reported by Professor Diener include:

- The happiest people tend to have strong social relationships.
- Working towards goals and achieving them are sources of well-being.
- Most people around the world, except those living in dire circumstances, report being happy the majority of the time, but very few report being consistently elated or extremely happy. Thus, slight to moderate happiness is the rule rather than the exception.
- However, people want not just to be happy, they want to be happy for the right reasons — for things they value. Happiness is thus a moral imperative, not simply a hedonistic one. Happiness results from people's values.
- Not only does happiness feel good, but happy people appear to function better than unhappy people — making more money, having better social relationships, being better organisational citizens at work, doing more volunteer work, and having better health. (From www.psych.uiuc.edu/~ediener/contributions_of_lab.htm, accessed 11 August 2007.)

Factors suggested to enhance midlife psychological well-being include:

*   good health
*   exercise
*   sense of control
*   positive social relationships
*   good marriage
*   mastery of multiple roles.

Parenting: Single Mothers and Psychological Well-being

A 2003 Irish study, *Family Well-being: What Makes a Difference?*, found that family type did not affect children's well-being negatively. Yet interestingly it did find that the psychological health of single mothers was impacted on negatively. According to the authors:

> As this variable [psychological well-being] is defined in terms of autonomy, environmental mastery, personal growth, positive relations with others, purpose in life and self-acceptance, it is quite plausible that single parents might have slightly lower scores, particularly given the inherent difficulties involved in raising children on one's own and in the context of the specific economic factors that often lead to single parenthood. Thus, the only effect of family type that has a statistically significant, robust impact on family well-being is the impact of raising children in a one-parent single family on the psychological well-being of mothers. (McKeown *et al.* 2003:70)

Thus it is the mothers, not the children, who are affected. The study found no evidence that family type impacts negatively on children's psychological well-being.

LATE ADULTHOOD

Late adulthood, the period from 65 years onwards, has been a neglected area in psychology. This is beginning to change, as more interest and research is being focused on this area, and there can be little doubt that the increasing longevity we are experiencing in the West is forcing us to reconsider this period of development. In Ireland a longitudinal study researching the lives of older people (TILDA) has been launched. Its stated aim is to gather information from a nationally representative survey which will become the foundation for policy formation and implementation to enable successful ageing in Ireland. The study will provide an accurate picture of the needs and contributions of older people.

An outline of this section:

- Physical development:
  - Physical changes
  - The health of the older population in Ireland
  - Ageing in Ireland
- Cognitive development:
  - Schaie's stages of cognitive development
  - Paul Baltes' dual-process model
- Socio-emotional development:
  - Erikson's integrity vs. despair
  - Levinson's season of man
- Ageism and the cohort effect
- The experiences of Irish elderly people.

Physical Development

Young-old people appear physically young for their actual age. Old-old people appear physically frail and show signs of decline.

Physical Changes

Some of the following changes are witnessed during old age. Individual variation must be taken into account as lifestyle and other factors can determine the rate of ageing and the magnitude of its effects.

- loss of brain weight increases after 60 years of age
- neurons (nerve cells) lost in visual, hearing and motor areas of the brain
- vision and hearing become increasingly impaired
- heart rate slows, as does blood flow
- sleep patterns change, and less sleep is needed
- appearance of skin continues to change, age spots develop
- loss of weight and height after 60
- muscle strength and flexibility decline.

The Health of the Older Population in Ireland

The *National Health Promotion Strategy 2000–2005* states:

> With the onset of middle age, Irish life expectancy figures begin to slip down the EU rankings and by the age of 65 years, life expectancy for both men and women is the lowest in the EU. Generally though we are living longer and it is projected that there will be a notable increase in the number of people over the age of 65 years by the year 2011–21. This is expected to pose a significant challenge to our health and social services. (Department of Health and Children 2000c:44)

SLAN 2002 data for respondents aged 65 years and over revealed that 38 per cent perceived their general health to be fair or poor. In relation to extreme/moderate problems experienced by older people, pain/discomfort and anxiety/depression rated highly. A cause for concern is that 36 per cent of older people surveyed reported taking no exercise and for those aged 75 years plus this rose to 51 per cent.

Twenty-one per cent of the 65–75 age group were current smokers and only 14 per cent in the over 75s smoked.

The challenge for health promotion is to improve longevity so that we live as long as our European counterparts. This can be achieved by promoting lifestyle changes, creating supportive environments and providing appropriate services for older people.

Ageing in Ireland

In 2007 the Central Statistics Office (CSO) produced a document entitled *Ageing in Ireland*, which outlines some key findings of people aged 65 and over in Ireland (p.10):

- In line with the increasing overall population, the number of persons aged 65 & over increased by 54,000 people between 1996 and 2006.
- In 2006, Ireland had the lowest proportion of its population aged 65 & over among EU countries at 11.0%, this compared to an EU 27 average of 16.8%.
- There is a projected upward trend in the 65 & over dependency ratios for both Ireland and the EU from 2006 to 2026. This dependency ratio is expected to increase from 16.4% to 25.1% for Ireland, and increase from 25.2% to 36.6% for the EU 25.
- The age specific death rate for males aged 65 & over has decreased from around 77 per 1,000 in 1980 to 51 in 2005. The corresponding decrease for

females was from 60 to 44 per 1,000 indicating a significant narrowing between both rates.

- In 2006, 29.5% of persons aged 65 & over indicated they had a disability compared to 9.3% of all persons. The proportion of persons with a disability increased with age, particularly for the older age groups. The disability rate varied from 18.7% for the 65–69 age group to 58.6% for the 85 & over age group.

- The proportion of women aged 65 & over living alone in Ireland (31.7%) was the eighth lowest of EU countries but the rate for men at 20.6% was the fourth highest.

- Ireland had the sixth highest employment rate for people aged 65 & over among EU countries in 2006. Men aged 65 & over in Ireland had a much higher rate of employment than the EU 27 average in 2006 (14.4% compared to 6.6%). The difference for women between Ireland (4.2%) and the EU 27 (2.8%) was less marked.

- In 2005, around 20% of persons aged 65 & over were at risk of poverty, which was substantially lower than the 2004 rate of 27.1%. This decrease was due mainly to an increase in the old age pension in 2005.

(CSO 2007:10)

This information should give you a perspective on the life experience of an older Irish citizen. Throughout this book, a constant theme has been the importance of placing a person within their context, and this is essential to understand the environment of the individual and to truly gain a greater knowledge of their developmental outcomes.

What do you think?
'There is a projected upward trend in the 65 & over dependency ratios for both Ireland and the EU from 2006 to 2026.' The CSO reports that the number of over 65s in Ireland will continue to grow in comparison to the rest of the population. What implications does this increasing 'dependency ratio' pose for government policy?

Cognitive Development

Schaie's Stages of Cognitive Development

- **Reorganisation** (end of middle adulthood to the beginning of late adulthood). This stage is marked by retirement. Cognitive tasks involve the reorganisation of life around new interests and pursuits.
- **Reintegrative** (late adulthood). Older adults look back over their lives, to make sense of life. The older person learns to be selective in using more limited energies.
- **Legacy creating** (advanced old age). This is the final stage of Schaie's theory and represents attempts by the older person to ensure that some part of them will continue after they are gone, that they leave a legacy. Activities such as making a will or funeral arrangements can be seen.

Check! Learning Goals

You should be able to:

- describe Schaie's final three stages
- evaluate his theory.

Paul Baltes Dual-process Model

Paul Baltes (1939–2006) was a leading researcher in the field of ageing and contributed enormously to our understanding of this area. Baltes proposed a dual-process model (1997) of cognitive functioning. It draws on the fluid and crystallised intelligences theory which we examined in relation to early adulthood. The dual-process model attempts to measure what aspects of intelligence are likely to improve or deteriorate with age.

Papalia *et al.* (2005:654) explain that in this model:

> ...**mechanics of intelligence** are the brain's neurophysiological 'hardware': information-processing and problem-solving function independent of any particular content. This dimension, like fluid intelligence, often declines with age. **Pragmatics of intelligence** are culture-based 'software': practical thinking, application of accumulated knowledge and skills, specialised expertise, professional productivity, and wisdom. This domain, which may continue to develop until very late adulthood, is similar to, but broader than, crystallised intelligence and includes information and know-how garnered from education, work, and life experience.

Check! Learning Goals

You should be able to:

- describe how the 'mechanics' and 'pragmatics' of intelligence differ
- identify which type of intelligence improves with age.

Socio-emotional Development

Erikson's Integrity vs. Despair

This is the final stage of Erikson's theory of psychosocial development. It is characterised by the features of integrity and despair. The former represents the positive outcome and is associated with a feeling of satisfaction with one's life and achievements. The individual feels content and at peace, reflecting, according to Erikson, psychosocial maturity on their part. However, if the outcome of this stage is despair the individual believes that they have made many mistakes in their life and that it is too late to correct them. They are unhappy looking back over their life and experience despair as they feel it is now too late to change things, which manifests itself in bitterness and anger. An individual in 'despair' may not be accepting of death.

Levinson's Season of Man

The Late Adult Transition (55 to 60)

As we saw in the previous section this is a period that lasts approximately five years while the individual makes the transition between middle and late adulthood.

The transition to late adulthood marks an opportunity to reflect upon successes and failures and enjoy the rest of life.

Check! Learning Goals

You should be able to:

- compare Levinson's and Erikson's stages
- explain the negative outcome that Erikson refers to and how it manifests itself.

Ageism and the Cohort Effect

In Chapter 2 we examined the theories of Glen H. Elder Jr, part of which referred to the 'cohort effect', meaning that lives can be influenced by social change. We compared the lives of people with disabilities born in the 1940s to those born today in order to elaborate on what is meant by a 'cohort effect'. It was clear that the lives of those born with a disability in Ireland today are better off than those born generations ago, though of course there is still some way to go before parity is reached.

If we apply the cohort effect to the older people, do you think their lives are better or worse than they were several generations ago? Today, older people live longer lives and generally have access to better healthcare services. However, in social terms, is the old adage of respecting your elders in as much evidence as it was many years ago? Many cultures have great respect for their older generation, for their wisdom and life experience. Yet in the West it appears that we have started to devalue older people. Current campaigns against ageism would seem to indicate that there are negative attitudes towards older people.

> What do you think? Have attitudes towards older people changed in the last 50 years? If so, why?

Check! Learning Goal
You should be able to:
- describe the 'cohort effect'.

The Experiences of Irish Elderly People

Horkan and Woods (1986) undertook research examining the lives and experiences of elderly people in suburban Dublin and highlighted some of the issues facing them. The report explores the social changes and difficulties experienced by older people. Through interviews with the elderly participants, four themes were identified:
- 'Life Has Changed'
- 'Life Can Be Lonely and Isolated'
- 'Life Can Be Dangerous'
- 'Life Can Be Difficult'.

The goal of the research was to highlight the issues facing older people and heighten public awareness of the problems they face. The authors argue that if older people are to continue living in the community we must be more aware of the issues facing them. (For more information visit www.ncaop.ie.)

You may wonder why there has been so much emphasis on physical health in a book covering developmental psychology. The relationship between physical and psychological well-being has been documented, but in addition to that you should be able to see the relationships that exist within an individual's environment, immediate and distant, and how these can influence their development. Recall Bronfenbrenner's ecological model in Chapter 2: to gain greater understanding of how we develop we must look at the different influences and how they interact to shape the developmental pathway. Bronfenbrenner further modified his theory to include the biological aspects of the individual, understanding this to be necessary to gain a fuller picture of the individual's development.

You should come away from this chapter with a greater understanding of how seemingly unrelated influences conspire in a holistic manner to shape how we develop. The choices we make in our 20s can shape the lives we have in our 60s. Food poverty, low income and other social variations that contribute to health outcomes point to the diverse influences that exist within Irish society. We also need to grasp the impact of societal attitudes and change on the lives of people as they age. Ageism is a real problem with many consequences for individual lives, yet we have seen that societal attitudes can be challenged and changed for the better. We all have our part to play.

In Focus: Alcoholism and Drug Addiction
What is Addiction?

Addiction, or dependence, is defined as 'a cluster of three of more symptoms listed below occurring at any time in the same 12-month period' (American Psychiatric Association 1994:176).

These symptoms are:

1. tolerance, or needing more and more of a substance to achieve the same effect
2. withdrawal, which involves unpleasant symptoms when the body is deprived of the substance, resulting in more frequent use to alleviate the negative symptoms
3. taking the substance for a longer period of time or in larger amounts than originally intended
4. unsuccessful desire to minimise use of the substance
5. much time spent to obtain, use, or recover from the effects of the substance
6. social, occupational or recreational activities are missed because of substance abuse
7. substance use is continued despite knowledge of causing a problem.

If neither tolerance nor withdrawal are present, then at least three of the remaining symptoms must be present (APA 1994).

Did You Know?
- One in 15 children will become alcoholic in their lifetime because of genetic predisposition (Weber & McCormick 1992).
- Children from alcoholic families are four to six times more likely to become alcoholic than children raised in non-alcoholic homes (ibid.).
- By their mid-20s, nearly 80 per cent of young adults have used an illicit drug.

There are two types of drug abuse: habitual and chronic polydrug abuse; and experimental or recreational abuse of drugs.

Carr (2003:590) remarks that there are seven categories of possible explanations for drug abuse:

- **First**, biological theories ‚of drug abuse focus on specific genetic factors; on temperamental attributes that are known to be strongly genetically determined; and on the role of physiological mechanisms in the development of tolerance, dependence and withdrawal.
- **Second**, intrapsychic deficit theories point to the importance of personal psychological vulnerabilities in the development of drug-using behaviour patterns.
- **Third**, cognitive-behavioural theories underlie the significance of certain learning processes in the genesis of drug problems.
- **Fourth**, family systems theories emphasise the importance of parental drug-using behaviour, parenting style and family organisation patterns in the aetiology and maintenance of drug abuse.
- **Fifth**, the role of societal factors such as social disadvantage, neighbourhood norms concerning drug use and abuse, and drug availability are the central concerns of sociological theories of drug abuse.
- **Sixth**, multiple-risk factor theories highlight the roles of factors at biological, psychological and social levels in the aetiology of drug abuse.
- **Finally**, change-process theories offer explanations for how recovery and relapse occur.

Impact of Drug Abuse on Interpersonal Adjustment
According to Carr, drug abuse impacts on interpersonal relationships. He outlines its possible ramifications:

> Within the family, drug abuse often leads to conflict. Individuals that abuse drugs within a peer-group situation may become deeply involved in a drug-oriented subculture and break ties with peers who do not abuse drugs. Some develop a solitary drug-using pattern and become more and more socially isolated as their drug using progresses ... Within the wider community, drug-related antisocial behaviour such as aggression, theft and selling drugs may bring youngsters into contact with the juvenile justice system ... Drug-related health problems and drug dependency may bring them into contact with the health service.
>
> (Carr 2003:589)

Can Addiction be Treated?
There are several successful treatments for addiction. It goes without saying that the earlier the addiction is treated, the easier it will be to control. Some of the most effective treatments for addiction are twelve-step programmes, such as the Alcoholics Anonymous system (Weber & McCormick 1992). Other treatments include individual and family therapy, group therapy, educational programmes, and self-esteem building and anger management workshops.

2006 National Report to EMCDDA
In Ireland each year a report is compiled and presented to the European Monitoring Centre for Drugs and Drug Addiction (EMCDDA). These are some of the findings of the 2006 report regarding the prevalence of drug use in Ireland:

- The results of the general population survey 2002/2003 indicate that one in five (18.5%) adults reported using an illegal drug in their lifetime. For young adults (aged 15–34 years) this rose to one in four (26.0%) people. Twice as many men as women reported the use of an illegal drug during the last month or the last year.
- The majority of those who have tried any illicit drug have used cannabis (marijuana or hashish). The lifetime prevalence rates for cannabis use are thus similar to those for use of any illicit drug and reflect the same trend. Lifetime

use of inhalants dropped slightly between 1999 (22%) and 2003 (18%) but remains high. The average for the 35 ESPAD countries in 2003 was 10%.

- The results of a national survey of third-level students were published in April 2005. Cannabis was the most common illicit drug used by students, with over one-third (37%) reporting that they had used it in the past 12 months. Ecstasy was the second most used illicit drug, followed by cocaine, magic mushrooms and amphetamines. For all drugs, the levels of use were higher among students than among those of a similar age group (15–24 years) in the general population. The use of solvents (inhalants) was particularly high.

Visit www.ndc.hrb.ie for more information on drug abuse in Ireland.

In Focus: Lifelong Physical Participation and its Benefits

Bill Creasey is a coach education officer and coach tutor who is actively involved in the progression and implementation of a national approach to physical fitness. He outlines the importance of lifelong activity to the health and well-being not just of the individual but for Irish society as well.

> While age is a relative thing, the 60-year-old man training for the Dublin City Marathon may be in better shape than the 20-year-old couch potato. However, generally speaking, after age 30 humans start losing muscle mass and after age 40 bone mass may decline.

Why do we Age?

Many theorists suggest that the human body simply wears out through the wear and tear of daily living. They compare the human body to an old car that breaks down and wears out parts through continued use and abuse. The old car ceases to perform and eventually stops. Although this may be the case with some human characteristics, it is not totally representative of the ageing process. On the contrary, investigations have shown that the use of the human body in exercise and physical activity can slow, stop, or in some cases reverse aspects of age-related deterioration.

The evidence is indisputable. Lifelong participation in physical activity has a significant positive impact on people's health and well-being. It is also evident that improved health and well-being have a significant positive consequence for both individuals and Irish society as a whole.

The Irish Sports Council (ISC) in its paper *Lifelong Involvement in Sport and Physical Activity* (LISPA) clearly emphasises the importance of lifelong physical activity in good health. It effectively includes all people, those with disabilities, the sedentary population and athletes involved in recreational and competitive sport who need to understand the health benefits of lifelong physical activity. The National Coaching and Training Centre (NCTC) state in their paper *Building Pathways in Irish Sport* that 'improving the health of the population enhances individual and social capital and thus supports economic and human development at local community level and for the country as a whole' (p.9). According to the National Coaching and Training Centre (2003), Shephard states:

> ... reports from our laboratory and elsewhere have suggested that a number of important social and economic benefits are associated with an increase of physical activity. Gains are seen in the workplace; greater productivity and reduced absenteeism, turnover and industrial injuries. Gains in the healthcare system include; fewer physician visits and less need for hospital utilization and geriatric care. Gains in the area of lifestyle include; reduction of appraised age and a lesser incidence of cigarette and alcohol abuse. Each of the Western nations might save billions of dollars if regular exercise were to be adopted by the entire population.

The table highlights the benefits of lifelong participation in physical activity.

Table 9.1: Benefits of lifelong participation in physical activity

Regular physical activity can:
- Increase muscular strength and endurance
- Increase aerobic fitness
- Increase flexibility
- Help control weight
- Decrease stress
- Increase feelings of well-being and self-esteem.

Beyond the many positive benefits associated with adequate physical activity, students need to know what diseases can be prevented by being active for life. Regular physical activity can reduce or prevent health problems such as:
- Premature death in general
- Death caused by heart disease
- Diabetes
- High blood pressure
- Some types of cancer
- High cholesterol levels.

NASPE (2005)

Recent technological and social changes have contributed to the development of a lifestyle in Ireland that is characterised by a sedentary way of life. The population has become increasingly overweight and obese and the health care system is beginning to shoulder the burden of hypokinetic diseases, that is, those relating to inactivity and unhealthy lifestyles. Deaths due to coronary heart disease in Ireland are the highest in Europe, while obesity has doubled among Irish men since 1990.

A sedentary lifestyle is a major element in poor health for a large number of Irish people. People who don't get enough physical activity are much more likely to develop health problems. Current statistics show that the average life expectancy of Irish people has increased.

Table 9.2: Life expectancy at birth (CSO 2007)

Year	Males	Females
1925–7	57.4	57.9
1965–7	68.6	72.9
2007	77.1	81.8

While continued improvements in the Irish healthcare system, lifestyle and environmental changes have resulted in a steady increase in life expectancy for men and women in Ireland, there is no room for complacency. Figures from SLAN indicate that significant numbers, especially in the older population, are not engaging in any form of physical activity.

Current physical guidelines (American College of Sports Medicine) advise that adults should engage in moderately intense exercise for at least 30 minutes, five days a week or vigorous exercise for at least 30 minutes three days a week. Also see the Irish Heart Foundation website for guidelines on Irish health promotion (www.irishheart.ie).

Queen's University, Belfast Research
Tully *et al.* (2007) carried out a 12-week study with adults aged 40 to 60 who were healthy but led sedentary lifestyles. According to the authors, the aim of the study was 'to determine, using unsupervised walking programmes, the effects of exercise at a level lower than currently recommended to improve cardiovascular risk factors and functional capacity.' Some participants (n=42) were given an exercise regime consisting of 30 minutes brisk walking five days a week. The other participants (n=40) were instructed to do 30 minutes brisk walking three days a week. The remaining 20 participants in the

study were told to maintain their normal lifestyles. Researchers found that there was a significant reduction in blood pressure and a slimming in size in both waist and hips, along with an increase in overall fitness. The non-walkers had no changes in any of these areas. The researchers conclude that the reduction in waist and hip circumference, and a reduction in blood pressure, can reduce the risk of dying from cardiovascular disease. Further they noted that the study shows that even moderate amounts of exercise (i.e. three times a week) can provide noticeable health benefits.

Table 9.3: Percentage doing no exercise at all in the week by gender, age and educational status

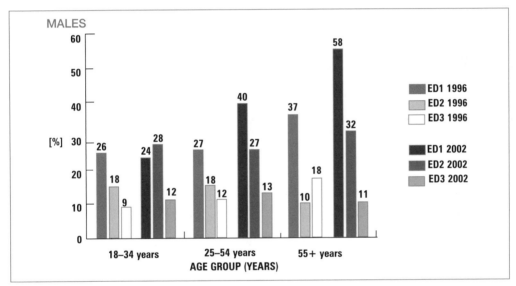

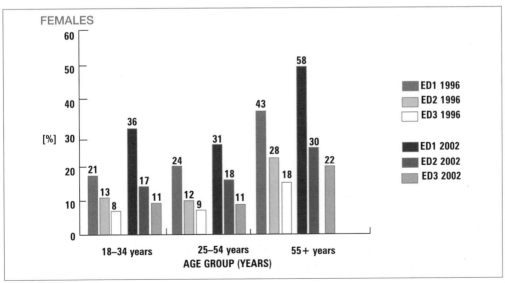

CHAPTER SUMMARY

- **Primary ageing** — age-related changes that reflect the biological aspects of ageing and which affect everyone.
- **Secondary ageing** — influenced by lifestyle and environmental factors.

Theories of ageing:

- damage-based — ageing results from a continuous process of damage which accumulates throughout the entire lifespan.
- programmed theories — ageing is predetermined and occurs on a fixed schedule.

Senescence is the deterioration of bodily functions that accompanies ageing in a living organism.

Early Adulthood

Physical Development

- second major brain spurt, continues until about 22 years of age
- 'wear and tear' approach — the body simply wears out over time.

SLAN findings:

- general health — majority report being in good health
- smoking — a decrease between 1998 and 2002, but younger people continue to smoke more than other age categories.

Cognitive Development

Schaie's stages of cognitive development:

- **acquisition** of information and skills in order to gain understanding of the world (childhood and adolescence)
- **achieving**: young adults use their knowledge to pursue a career, choose a lifestyle (late teens and early adulthood).

Postformal stage: a fifth stage after Piaget's formal operational stage. We are able to apply this new level of thinking to problem-solving and reasoning. Requires reflective thinking.

Theory of fluid and crystallised intelligence (Cattell & Horn) proposed that general intelligence is roughly 100 abilities working together.

- **fluid intelligence** drives individual's ability to think and act quickly, peaks in early adulthood and then declines
- **crystallised intelligence** stems from learning, cultural influences and personality factors, continues to improve as individuals age.

Socio-emotional Development

Erikson's isolation vs. intimacy (20–30):
- **intimacy** — positive — relates to forming an intimate or love relationship and committing to it
- **isolation** — negative — reflects an individual's hesitation to form a relationship, possibly due to a fear of losing identity or to self-absorption.

Levinson's seasons of man:
- nine developmental periods with four 'seasonal cycles'
- **life structure** — underlying pattern of an individual's life at a given time; shaped by their social and physical environment
- **life cycle** — underlying order spanning adult life, consisting of a sequence of eras. The move from one era to another is not smooth: there are cross-era **periods of transition**, which last for about five years.

Early adulthood stages:
- **Early Adult Transition (17–22)** — developmental bridge between pre-adulthood and early adulthood.
- **Entry Life Structure for Early Adulthood (22–28)** — building and maintaining an initial mode of adult living.
- **Age 30 Transition (28–33)** — opportunity to reappraise and modify entry structure and create basis for next life structure.
- **Culminating Life Structure for Early Adulthood (33–40)** — the vehicle for completing this era and realising youthful ambitions.
- **Midlife Transition (40–45)** — another cross-era shift; terminates early adulthood and initiates middle adulthood. Main focus of this stage is career and family.

Middle Adulthood
Physical Development
- hearing loss for high-pitched sounds

- ability to focus declines, near vision declines
- motor functioning decreases
- strength and co-ordination decline
- loss of muscle mass, replaced by fat

Menopause = end of reproductive capacity, cessation of periods.

Cognitive Development

Schaie's stages of cognitive development:
- **responsibility** — individuals use their abilities to solve problems related to their responsibilities to others, e.g. family members.
- **executive** — reflects a concern for the welfare of the broader social system. People deal with complex relationships on multiple levels.

Sternberg's triarchic theory of successful intelligence: intelligent behaviour is reflected in optimal adaptation to your environment or particular sociocultural context and involves the amalgamation of three abilities:
- **analytical** — enable the individual to evaluate, analyse, compare and contrast information
- **creative** — generate invention, discovery, and other creative endeavours
- **practical** — tie everything together by allowing individuals to apply what they have learned in the appropriate setting.

To be successful the individual should pick an environment best suited to their individual abilities.

Socio-emotional Development

Erikson's generativity vs. stagnation:
- **generativity** — offering the opportunity to the individual to reach out to and guide others; focus is beyond the self and towards others
- **stagnation** — negative outcome — occurs when the person places their needs above challenge and sacrifice.

Levinson's middle adulthood stages:
- **Midlife Transition (40–45)** — cross-era shift: terminates early adulthood and initiates middle adulthood.

- **Entry Life Structure for Middle Adulthood (45–50)** — provides an initial basis for life in a new era.
- **Age 50 Transition (50–55)** — opportunity for improving the entry life structure.
- **Culminating Life Structure for Middle Adulthood (55–60)** — the framework in which we conclude this era.
- **Late Adult Transition (55–60)** — a boundary period between middle and late adulthood, separating and linking the two eras.

Factors in midlife psychological well-being include:
- good health
- exercise
- sense of control
- positive social relationships
- good marriage.

Late Adulthood

Late adulthood = the period from 65 years onwards.

Physical Development
- young-old people appear physically young for their age
- old-old people appear physically frail and show signs of decline.

Physical changes:
- loss of brain weight increases after 60 years of age
- neurons (nerve cells) lost in visual, hearing and motor areas of the brain
- heart rate slows, as does blood flow
- sleep patterns change, less sleep needed.

Life expectancy in Ireland for both men and women is the lowest in the EU.

SLAN:
- 38 per cent perceived their general health as fair or poor
- 36 per cent reported taking no exercise
- 51 per cent aged over 75 years took no exercise.

CSO 2007, *Ageing in Ireland*:

- In line with the increasing overall population, the number of persons aged 65 and over increased by 54,000 between 1996 and 2006.
- In 2006, the proportion of Ireland's population aged 65 and over was the lowest among EU countries: 11 per cent as compared to an EU27 average of 16.8 per cent.

Cognitive Development

Schaie's stages of cognitive development:

- **reorganisation** — retirement and reorganisation of life around new pursuits
- **reintegrative** — older adults look back over their lives, to make sense of life
- **legacy creating** — advanced old age. Final stage of Schaie's theory; represents attempts by the older person to leave a legacy.

Paul Baltes' dual-process model draws on the fluid and crystallised intelligences theory, and attempts to measure what aspects of intelligence are likely to improve or deteriorate with age.

- **Mechanics of intelligence** — the brain's 'hardware', physiologically based and involved in information-processing and problem-solving, declines with age.
- **Pragmatics of intelligence** — culture-based 'software', practical thinking, expertise and wisdom, continues to develop until very late adulthood.

Socio-emotional Development

Erikson's integrity vs. despair — the final stage of Erikson's theory of psychosocial development.

- **integrity** — positive outcome — associated with a feeling of satisfaction with one's life and achievements
- **despair** — negative — individual believes they have made many mistakes in their life and that it is too late to correct them. An individual in 'despair' may not be accepting of death.

Levinson's season of man:

- **Late Adult Transition (55–60)** — the transition to late adulthood, opportunity to reflect upon successes and failures and enjoy the rest of life.

Ageism and the cohort effect: the 'cohort effect' refers to the way lives can be influenced by social change, change in attitudes towards older people.

CHAPTER TEN

DEATH AND DYING

This is a difficult chapter for most people as many of us have had experience of personal loss of a loved one. This chapter is not intended to offer advice or to instruct on how to 'counsel' a person who is dying or an individual who has lost a loved one. Its purpose is to introduce some psychological explanations of how we make sense of or cope with dying. The main focus will be from the perspective of the person who is dying rather than those surrounding them. Several approaches will be examined, and children's understanding of death is also explored. The chapter ends with Murray Parkes's (1998) discussion of bereavement reactions. Robert Kastenbaum's 2000 book *The Psychology of Death* is referred to extensively in this chapter as it is a particularly comprehensive reference book on all psychological aspects of death and dying.

CHAPTER OUTLINE

- Terminology:
  - Phases of dying
- Leading causes of death in Ireland
- Models of dying:
  - Perspectives
  - Awareness and interaction
  - Kübler-Ross stages of dying
  - The existential perspective
  - Coping with dying as a developmental task
- Children's understanding of death:
  - Nagy's drawings
- Reactions to bereavement:
  - The trauma response
  - The grief response
  - The psychosocial response
- The last word

TERMINOLOGY

Kastenbaum (2000) remarks that the following terminology has evolved because more direct terms such as 'death' and 'dying' are '... too plain spoken ... for some tastes' (p. 200).

- **life-threatening condition** may be used if there is some possibility for recovery
- **end of life issues** refers to any decisions regarding the end of life, from medical directives to funeral arrangements.

Phases of Dying

- **Agonal phase** — derived from the Greek word *agon,* meaning struggle. Muscles may jerk and breathing becomes difficult as the body can no longer support life.
- **Clinical death** — breathing and heartbeat have ceased, but resuscitation may be possible.
- **Death** occurs when all activity in the brain and brain stem has stopped, resulting in brain death, which is irreversible.

LEADING CAUSES OF DEATH IN IRELAND

The Central Statistics Office (CSO 2006b) reported that almost four in every five deaths were from diseases of the circulatory system (35 per cent), cancer (30 per cent), or diseases of the respiratory system (13 per cent). They reported a death rate of 6.3 per 1,000 population.

MODELS OF DYING

Perspectives

The **personal perspective** acknowledges that each of us has our own views on dying and death. This perspective is shaped by our life experiences, family, cultural and religious beliefs and even the mass media. Our personal perspective influences our reactions and perceptions of death.

In the **role-specific perspective**, Kastenbaum remarks that there are '... perspectives on dying that are influenced by education, role, and responsibility ... the clergyperson's personal perspective on dying is expected to function within the established framework of theology and custom' (2000:210). Others with this perspective include nurses, doctors and trained counsellors, who have specific roles in the process of death and dying. They are influenced by their role and its responsibilities.

The four main models of dying are:

- awareness and interaction
- Kübler-Ross stages of dying
- coping with dying as a developmental task
- the existential perspective.

Awareness and Interaction

David Sudnow (1967), in *Passing On: The Social Organisation of Dying*, found that the attention people received in life or death situations was '... significantly influenced by hospital staff perception of their age, present ability and social situation' (cited in Kastenbaum 2000:212). Discrimination and prejudice continue right until the end of life, and this highlights the effect of socio-economic factors in influencing an individual's outcome.

Barney Glaser and Anselm Strauss (1966, 1968) wrote two pieces based on their observations of interactions in several hospitals. From these observations they determined that different types of 'awareness and interaction' patterns exist. 'Awareness and interaction' refers to awareness on the part of the patient and the communication (interaction) pattern with another (family or medical professional).

Four basic types were identified:

Closed awareness:

The dying person does not realise that he or she is dying and the careprovider or the interaction partner is not going to tell them. This is a keeping-the-secret context of the communication (or, rather, noncommunication). It is of great psychological interest because [of] the tension involved in trying to interact positively with a dying person while at the same time avoiding a disclosure that one fears would be devastating.

Suspected awareness:

The patient surmises that he or she is not being told the truth, all the truth, and nothing but the truth. There is an effort to check out one's growing belief that the progress of the disease cannot be reversed or halted. The other interactant may or may not suspect that the patient is suspecting, but in any case is resolved not to let the truth slip out.

Open awareness:

Both the patient and others are willing and able to share their knowledge and concerns. This does not necessarily mean that they have extensive discussions of dying and death. Rather, it means that everybody involved feels free to bring up the subject whenever it seems necessary or useful to do so. The open awareness context is almost invariably accompanied by less tension and conflict than either of the first two noted.

Mutual pretense:

This is perhaps the most subtle and certainly the most discussed type of interaction context. The patient knows he or she is dying. The visitor knows the patient is dying. Both pretend that they don't know. Furthermore, both may be aware that the other person is pretending, too. The mutual pretense act is difficult to sustain; one or the other person may be caught off guard or under pressure and let the horrifying truth escape.

(Cited in Kastenbaum 2000:214–15)

Kübler-Ross Stages of Dying

Elisabeth Kübler-Ross's book *On Death and Dying*, first published in 1969, proved to be a seminal one. This area had been ignored and Kübler-Ross used her experience working as a doctor in American hospitals to shine a light on it. Kübler-Ross noted that there are a number of differences in how different cultures treat death. She recounted how as a child in her homeland (Switzerland) she and other neighbours visited a local farmer as he lay dying at home and noted that dying and death seemed to be far more 'natural' than the medicalised environment and approach that appeared to dominate in American hospitals. She believed that a fear of death was strong in American culture and may be due to the fact that the culture had made death cold, sterile, impersonal and mechanised. Kübler-Ross noted that the American view of death is abhorrence and avoidance of death, whereas in other cultures it is treated as part of a normal transition.

Based on her observations and work with dying patients she devised the following five stages of dying that are passed through:

1 **Denial and isolation** — 'No, not me, it cannot be true.' The first response is a temporary state of shock. This stage is characterised by isolation: sometimes dying

people withdraw. Others occasionally withdraw from them.

2   **Anger** — 'Life is not fair! Why me?' When denial can no longer be maintained it is replaced by anger, rage, envy and resentment. The anger may be displaced in all directions, the person may feel a loss of control. Life is a constant reminder of death.

3   **Bargaining** — 'If you'll just let me live until my grandchild is born ...' This appears to be seeking the chance that one will be rewarded for good behaviour, but really it represents an attempt to postpone the death.

4   **Depression** — 'What difference does it make anyway?' Rage and anger are displaced by a sense of loss at the inevitable: this may represent a preparatory grief as the person prepares for death. Other behaviours that might be seen are reactive depression, loss of self-esteem and guilt.

5   **Acceptance** — 'I've lived a good life, I'm ready to go.' If the patient has enough time, they will become neither depressed nor angry, happy nor sad, but void of feelings ('all felt out'). Dying patients find peace and acceptance, their circle of interests diminish and they prefer fewer visits and distractions.

Individual Differences

Some patients fight to the end, others just give up. The best forms of support for dying patients include encouragement, maintenance of dignity and independence.

Criticisms

*   Not all individuals go through the same sequence.
*   Some people struggle to the end.

Homer Simpson and Kübler-Ross

On a somewhat lighter note, the Kübler-Ross stages of dying are depicted in *The Simpsons* when Homer learns that he is dying. This is a transcript (by Rick Froman) of the sketch from the eleventh episode in the second series, entitled *One Fish, Two Fish, Blowfish, Blue Fish*.

Dr Hibbert: Now, a little death anxiety is normal. You can expect to go through five stages. The first is denial.
Homer: No way! Because I'm not dying! [hugs Marge]
Dr Hibbert: The second is anger.
Homer: Why you little! [steps towards Dr Hibbert]

Dr Hibbert: After that comes fear.

Homer: What's after fear? What's after fear? [cringes]

Dr Hibbert: Bargaining.

Homer: Doc, you gotta get me out of this! I'll make it worth your while!

Dr Hibbert: Finally, acceptance.

Homer: Well, we all gotta go sometime.

Dr Hibbert: Mr Simpson, your progress astounds me.

(www.intropsychresources.com/pmwiki/pmwiki/pmwiki.php?n=ResourcesByType.Cartoons)

The Existential Perspective

One of the most influential books I've read is Ernest Becker's *The Denial of Death* (1973), which won the Pulitzer Prize in 1974. Its basic tenet is that humans spend most of their lives in terror of death and denying the reality of it. Several themes emerge in his writing. First, the world is terrifying and the basic motivation for human behaviour is our biological need to control our basic anxiety, to deny the terror of death. Since terror of death is so overwhelming we conspire to keep it unconscious. The issue of meaning is central to the existential perspective of death and also of the life lived. A central question is: what is the meaning of life if death is meaningless?

Coping with Dying as a Developmental Task

A lifespan approach recognises that as we progress through our lives we meet a series of developmental challenges. In this light, dying is seen as perhaps the final task that awaits us. Charles A. Corr suggests that four challenges await the dying person: physical, psychological, social and spiritual. Doka (1993) depicts 'phase-specific' tasks that face the individual from pre-diagnosis to recovery (should that occur).

According to Kastenbaum (2000:220) the Corr/Doka perspective offers several important guidelines:

- Dying people still have a need to accomplish. One does not cease coping and striving; indeed, one calls upon all available resources to accomplish the final set of tasks.
- The lifespan developmental task encourages attention to a very broad range of problems and thereby reduces the temptation to view the dying process within a narrowed and possibly oversimplified perspective.
- The idea of having to achieve something when we are ageing, and again when we are dying, seems to resonate well with the product- and goal-directed character of our

society. People who have become achievement- and success-oriented throughout their lives may find it logical and meaningful to apply this approach to the very end of live.

CHILDREN'S UNDERSTANDING OF DEATH

Nagy's Drawings

Maria Nagy conducted research with children with the aim of understanding how they conceive of death. To do this she asked the children (ranging in age from three to ten) to draw pictures representing their ideas of death and to explain their pictures. From this Nagy was able to construct a developmental picture of children's conception of death.

The three phases were:

Stage One: 0–5 Years Old

A child of this age does not recognise that death is final. The child looks upon death as being continuous with life; the dead person has merely gone away. It is the separation element that is most distressing to the child. Children of this age are 'egocentric' in terms of cognitive thinking (see Chapter 6) and this is reflected in their understanding of death. Kastenbaum relates, '… much of the "deadness" comes from the deceased's invisibility to and distance from one's [the child's] self. Furthermore the dead are understood with reference to one's own experiences (e.g., they don't get as hungry as I do).' (ibid., 52).

Stage Two: 5–9 Years Old

The child begins to understand that death is final. 'Death personifications' begin to dominate during this phase, with death represented as a figure or person. Two images feature strongly: 'Death' as a separate person and 'Death' depicted as a dead person. 'Death' was usually male and was depicted as a skeleton, angel or even a clown. A theme that was present was that if you were smart enough or ran fast enough you could avoid the death-man.

Stage Three: 9–10 Years Old

These children recognised the finality of death as well as its universality, that it comes to all people eventually. It would appear that the child has attained an adult understanding of death.

REACTIONS TO BEREAVEMENT

Colin Murray Parkes (1998:200) outlines three responses common to those who have been bereaved.

The Trauma Response

- An alarm reaction — anxiety, restlessness, and the physiological accompaniments of fear.
- Anger and guilt, including outbursts directed against those who press the bereaved person towards premature acceptance of the loss.
- Post-traumatic stress disorder: either the specific pathological response or the less specific anxiety states and panic syndromes.

The Grief Response

- An urge to search for and to find the lost person in some form.
- Relocating the lost person, including identification phenomena –- the adoption of traits, mannerisms or symptoms of the lost person, with or without a sense of that person's presence within the self.
- Pathological variants of grief: the reaction may be excessive and prolonged or uninhibited and inclined to emerge in distorted form.

The Psychosocial Response

- A sense of dislocation between the world that is and the world that should be, often expressed as a sense of mutilation or emptiness, which reflects the individual's need to relearn their internal model of the world.
- A process of realisation, i.e. the way in which the bereaved moves from denial or avoidance of recognition of the loss towards acceptance and the adoption of a new model of the world.
- This process may be impaired by the feelings of helplessness and hopelessness that characterise depression.

THE LAST WORD

John Diamond, writer and columnist, charted his battle with cancer in his column in *The Times* and in a television documentary. He also wrote a book called *C: Because Cowards Get Cancer, Too*. In the following extract he has just received the news that the cancer is terminal. John Diamond died aged 47 in March 2001, leaving a widow (Nigella Lawson) and two children.

I'd imagined that I'd feel terrified when I got the news, but what I felt most of all was sad. Sad for Nigella, the children, my parents. As if, of course, sad were a word up to this particular job. I realised that the reason I don't seem to be going through the standard denial-anger-bargaining with God-acceptance schtick is because that's what I've been doing for the past 20 months or so. As soon as I heard the first diagnosis I heard a death sentence being passed and I suppose I never thought of the various operations and procedures as much more than temporary reprieves. Living with cancer must always mean living with the threat of death, even, I imagine, if you manage to increase the distance between you and the diagnosis to the five years that counts as a cure.

Meanwhile, I have some affairs to get in what passes for order. We haven't told the children and won't for a while, so if you come across them, please don't say anything. I'll carry on working as long as I can and given that one side-effect of the chemotherapy is fatigue, I'm sure you'll understand if I don't answer your mail individually from now on. And we're planning a big party to celebrate Nigella and my ten years of being together. It's strange how, in the middle of all this madness, there are some things worth celebrating.

And so this is how you find me. Not quite waiting to die, because although I've accepted that I will, and sooner rather than later, the same rules apply to the foreshortened life as to one of normal length: just as no well-balanced 45-year-old says 'Why bother going to the movies? I'll be dead in 30 years,' so I find that my imminent death doesn't stop me wanting to know what happens at the end of bad detective thrillers or wanting to spend time with Nigella and the children. Those things are still worth doing. As I write this we have all just returned from buying a basket for the spaniel we are to collect in a couple of days' time. A friend e-mailed me when she heard this to tell me it's a denial of what's happening and what's about to happen. It isn't: I know what's happening. But a dog is a happy thing, and it will be happy for me for whatever time I've got left and as happy as things can be for the family when I've gone.

(www.times-archive.co.uk/JohnDiamond/)

This brings us to the end of our journey across the lifespan, of which only a glimmer has been covered in this book, but hopefully it has left you with more questions than it has answered. Good luck on your journey wherever your path may take you!

BIBLIOGRAPHY

Ainsworth, M. D. S., Blehar, M. C., Waters, E. and Wall, S. (1978). *Patterns of Attachment: A Psychological Study of the Strange Situation*. Hillsdale, NJ: Erlbaum.

American Psychiatric Association (APA) (1994). *Diagnostic and Statistical Manual of Mental Disorders* (4th edn). Washington, DC: American Psychiatric Press.

Andrews, A., Ben-Arieh, A., Carlson, M., Damon, W., Dweck, C., Earls, F., Garcia-Coll, C., Gold, R., Halfon, N., Hart, R., Lerner, R. M., McEwen, B., Meaney, M., Offord, D., Patrick, D., Peck, M., Trickett, B., Weisner, T. and Zuckerman, B. (Ecology Working Group) (2002). *Ecology of Child Well-Being: Advancing the Science and the Science-Practice Link*. Georgia: Centre for Child Well-Being.

Aslin, R. N. (1987). 'Motor aspects of visual development in infancy' in N. P. Salapatek and L. Cohen (eds), *Handbook of Infant Perception, Vol.1: From sensation to perception* (pp 43–113). Orlando: Academic Press.

Bandura, A., Ross, D. and Ross, S. A. (1961). 'Transmission of aggression through imitation of aggressive models', *Journal of Abnormal and Social Psychology*, 63, 575–82.

Baumrind, D. (1966). 'Effects of authoritative parental control on child behavior', *Child Development*, 37(4), 887–907.

Beck, A. T. (1976). *Cognitive Therapy of Depression*. Guildford Press.

Becker, E. (1997). *The Denial of Death*. New York: Free Press.

Begley, M., Chambers, D., Corcoran, P. and Gallagher, J. (2004). *The Male Perspective: Young Men's Outlook on Life*, National Suicide Research Foundation. Retrieved from www.nsrf.ie/reports/CompletedStudies/YoungMensStudy.pdf (accessed 4 July 2007).

Belsky, J., Steinberg, L. and Draper, P. (1991). 'Childhood experience, interpersonal development, and reproductive strategy: an evolutionary theory of socialization', *Child Development*, 62, 647–70.

Benson, N. (2003). *Introducing Psychology*. England: Icon Books.

Blyth, D. and Traeger, C. (1988). 'Adolescent self-esteem and perceived relationships with parents and peers', in S. Selzinger, J. Antrobus and M. Hammer (eds), *Social Networks of Children, Adolescents and College Students*. Hillsdale, NJ: Erlbaum Associates.

Bowlby, J. (1951) 'Maternal care and mental health', World Health Organization monograph (Serial No. 2).

Boyd, D. and Bee, H. (2005). *Lifespan Development* (4th edn). New York: Allyn & Bacon.

Bretherton, I. (1992). 'The origins of attachment theory', *Developmental Psychology*, 28, 759–75.

Bronfenbrenner, U. (1979). *The Ecology of Human Development*. Cambridge, MA: Harvard University Press.

Brooks-Gunn, J., Petersen, A. C. and Eichorn, D. (1985). 'The study of maturational timing effects in adolescence', *Journal of Youth and Adolescence*, 14(3), 149–61.

Bruch, H., Czyzewski D. and Suhr, M. A. (1988). *Conversations with Anorexics*. Basic Books.

Buck, R., and Ginsburg, B. E. (1991). 'Emotional communication and altruism: the communicative gene hypothesis', in M. Clark (ed.), *Altruism: Review of Personality and Social Psychology*, vol. 12 (pp. 149–75). Newbury Park, CA: Sage Publications.

Bushnell, I. W. R., Sai, F., and Mullin, J. T. (1989). 'Neonatal recognition of the mother's face', *British Journal of Developmental Psychology*, 7, 3–15.

Buss, A. H. and Plomin, R. (1984). *Temperament: Early Developing Personality Traits*. Hillsdale, NJ: Erlbaum.

Caplan, M. Z. and Hay, D. F. (1989). 'Preschoolers' responses to peers' distress and beliefs about bystander intervention', *Journal of Child Psychology and Psychiatry and Allied Disciplines*, 30(2), 231–42.

Carr, A. (2003). *The Handbook of Clinical and Adolescent Clinical Psychology: A Contextual Approach*. New York: Brunner-Rutledge.

Central Statistics Office (CSO) (2006a). *Women and Men in Ireland*. Dublin: Stationery Office.

—. (2006b). *Vital Statistics: Fourth Quarter and Yearly Summary*. Retrieved from www.cso.ie/releasespublications/documents/vitalstats/current/vstats.pdf (accessed 11 July 2007).

—. (2007). *Ageing in Ireland*. Dublin: Stationery Office.

Chomsky, N., Hauser, M. and Tecumseh Fitch, W. (2002). 'The faculty of language: what it is, who has it, and how did it evolve?', *Science*, 298, 1569–79.

Cianciolo, A. T. and Sternberg, R. J. (2004). *Intelligence: A Brief History*. Malden, MA: Blackwell.

Clarke, A. M. and Clarke, A. D. B. (1976). *Early Experiences: Myth and Experience*. London: Open Books.

—. (1999). 'Early experience and the life path', in S. Ceci and W. Williams (eds), *The Nature–Nurture Debate: The Essential Readings*. Oxford: Blackwell.

Cohen, D. and Volkmar, F. (1997). *Autism and Pervasive Developmental Disorders* (2nd edn). New York: John Wiley & Sons.

Colby, A., Kohlberg, L., Gibbs, J. and Lieberman, M. (1983). 'A longitudinal study of moral judgment', *Monographs of the Society for Research in Child Development*, 48 (1–2), serial no. 200.

Crain, W. C. (1985). *Theories of Development: Concepts and Applications*. London: Prentice-Hall.

Crockett, L. and Petersen, A. (1987). 'Pubertal status and psychological development: findings from early adolescence studies', in Lerner and Foch (eds), *Biological-Psychosocial Interactions in Early Adolescence: A Lifespan Perspective*. Hillsdale NJ: Erlbaum.

Croft, C., O'Connor, T. G., Keaveney, L., Groothues, C., Rutter, M. and the ERA team (2001). 'Longitudinal change in parenting associated with developmental delay and catch-up', *Journal of Child Psychology and Psychiatry*, 42(5), 649–59.

Curtiss, S. (1977). *Genie: A Psycholinguistic Study of a Modern Day 'Wild Child'*. London: Academic Press.

Deacon, T. (1997). *The Symbolic Species*. London: Allen Lane.

DeCasper, A. J. and Spence, M. J. (1986). 'Prenatal maternal speech influences newborns' perception of speech sounds', *Infant Behaviour and Development*, 9, 133–50.

Dekovic, M. and Janssens, J. (1992). 'Parents' rearing style and child's sociometric status', *Developmental Psychology*, 28, 925–32.

Department of Health and Children (2000a). *EHLASS Report for Ireland 2000*. Dublin: Stationery Office. Retrieved from: http://ec.europa.eu/health/ph_projects/2000/injury/fp_injury_2000_frep_19_en.pdf (accessed 10 June 2007).

—. (2000b). *The National Children's Strategy: Our Children – Their Lives*. Dublin: Stationery Office.

—. (2000c). *National Health Promotion Strategy 2000–2005*. Dublin: Stationery Office. Retrieved from www.dohc.ie/publications/pdf/hpstrat.pdf?direct=1.

—. (2002a). *Traveller Health: A National Strategy, 2002–2005*. Retrieved from www.dohc.ie/publications/traveller_health_a_national_strategy_2002_2005.htm (accessed 14 July 2007).

—. (2002b). *Health Statistics 2002*. Dublin: Stationery Office.

—. (2002c, 2006). *Health Behaviour in School-aged Children (HBSC)*. Dublin: Stationery Office.

—. (2005). *Report on Perinatal Statistics 2002*, ESRI. Retrieved from: www.dohc.ie/publications/perinatal_statistics_2002.html (accessed 1 August 2007).

Department of Health and Children and Centre for Health Promotion Studies, NUI Galway (1998, 2002). *Survey of Lifestyle, Attitudes and Nutrition (SLAN)*. Retrieved from www.nuigalway.ie/health_promotion/documents/slan03.pdf (accessed 10 July 2007).

De Wolff, M. S. and Van Ijzendoorn, M. H. (1997). 'Sensitivity and attachment: a meta-analysis on parental antecedents of infant attachment', *Child Development*, 68, 571–91.

Diamond, J. (1999). *C: Because Cowards get Cancer Too*. London: Vermilion.

Dubas, J. S., Graber, J. A. and Petersen, A. C. (1991). 'The effects of pubertal development on achievement during adolescence', *American Journal of Education*, 99, 444–60.

Eisenberg, N. and Mussen, P. (1989). *The Roots of Pro-social Behaviour in Children*. Cambridge: Cambridge University Press.

Elder, G. H. (2001). 'Human lives in changing societies: life course and developmental insights', in R. Cairns, G. H. Elder and E. J. Costello (eds), *Developmental Science*. Cambridge: Cambridge University Press.

Elkind, D. (1967). 'Egocentrism in adolescence', *Child Development*, 38, 1025–33.

Ellis, B. J. and Garber, J. (2000). 'Psychosocial antecedents of variation in girls' pubertal timing: maternal depression, stepfather presence, and marital and family stress', *Child Development*, 71, 485–501. Retrieved from http://ag.arizona.edu/fcs/fshd/people/ellis/CD%20Ellis%20&%20Garber%202000.pdf (accessed 17 August 2007)

Erikson, E. H. (1963). *Childhood and Society* (2nd edn), New York: Norton.

—. (1980). *Identity and the Lifecycle*. New York: Norton.

Eron, L. D., Walder, L. O. and Lefkowitz, M. M. (1971). *Learning of Aggression in Children*. Boston: Little, Brown.

Fantz, R. L. (1961). 'The origin of form perception', *Scientific American*, 204, No.5, 66–72.

Farrington, D. P.(1991). 'Childhood aggression and adult violence: early precursors and later outcomes', in D. J. Pepler and K. H. Rubin (eds), *The Development and Treatment of Childhood Aggression* (pp 5–30). Hillsdale, NJ: Erlbaum.

Field, H. and Domangue, B. (1987). *Eating Disorders Throughout the Lifespan*. Greenwood Press.

Freud, S. (1964). 'An outline of psychoanalysis', in J. Strachey (ed. & trans.), *The Standard Edition of the Complete Psychological Works of Sigmund Freud* (vol. 22). London: Hogarth.

Frith, U. (1996). *Autism: Explaining the Enigma*. Oxford: Blackwell Press.

Garcia-Moro, C. and Hernandez, M. (1990). 'Changes in age at menarche in Spain, 1909–1965', *International Journal of Anthropology*, 5(2), 117–24.

Gardner, H. (1983). *Frames of Mind: The Theory of Multiple Intelligence*. New York: Basic Books.

Garhart Mooney, C. (2000). *Theories of Childhood: An Introduction to Dewey, Montessori, Erickson, Piaget and Vygotsky*. Minnesota: Redleaf Press.

Garnefski, N. and Arends, E. (1998). 'Sexual abuse and adolescent maladjustment: differences between male and female victims', *Journal of Adolescence*, 21, 99–107.

Gibson, E. J. and Walk, R. D. (1960). 'The "visual cliff"', *Scientific American*, 202, 64–71.

Giedd, J. (2004). 'What makes teens tick?', *Time*, 10 May 2004.

Greene, S. (2007). ' Children's recovery after early adversity: Lessons from inter-country adoption'. Retrieved from www.tcd.ie/childrensresearchcentre/index.php?id=129 (accessed 11 August 2007).

Griffin, S. and Shevlin, M. (2007). *Responding to Special Educational Needs: An Irish Perspective*. Dublin: Gill and Macmillan.

Hall, C. (1962). *A Primer of Freudian Psychology*, London.

Hall, D. (1985). 'Technical note: extreme deprivation in early childhood', *Journal of Child Psychology and Psychiatry*, 26(5), 825.

Harlow, H. F. (1961). 'The development of affectional patterns in infant monkeys', in B. M. Foss (ed.), *Determinants of Infant Behaviour* (pp 75–97). London: Methuen.

Harter, S. (1990). 'Causes, correlates and the functional role of global self-worth: a life-span perspective', in R. J. Sternberg and J. Kolligian (eds), *Competence Considered*. New Haven: Yale.

Hartup, W. W. (1970) 'Peer interaction and social organization' in P. H. Mussen (ed.), *Carmichael's Manual of Child Psychology, Vol. 2* (3rd edn). New York: Wiley.

Hay, D. F. (1994). 'Prosocial development', *Journal of Child Psychology and Psychiatry*, 35, 29–71.

Hayes, N. (2005), *Early Childhood: An Introductory Text* (3rd edn). Dublin: Gill and Macmillan.

Hayflick, L. (1980). 'The cell biology of human aging', *Scientific American*, 242, 56–65.

Herbert, M. (2002) *Typical and Atypical Development: From Conception to Adolescence*. Oxford: BPS Blackwell.

Hetherington, E. M. and Parke, R. D. (1999). *Child Psychology: A Contemporary Viewpoint* (5th edn). New York: McGraw-Hill.

Hoffman, M. (1990). 'The origins and development of empathy', *Motivation and Emotion*, 14 (2), 75–80.

Honigmann, I. and Honigmann, J. (1954). 'Child rearing patterns among the Great Whale River Eskimo', in H. R. Schaffer (1996), *Social Development*. Oxford: Blackwell.

Horkan, M. and Woods, A. (1986). *This is Our World: Perspectives of Some Elderly People on Life in Suburban Dublin*. Report No. 12, National Council for the Aged. Retrieved from www.ncaop.ie/publications/research/reports/This_Is_Our_World12.pdf (accessed 11 September 2007).

Horn, J. L. and Cattell, R. B. (1967). 'Age differences in fluid and crystallized intelligence', *Acta Psychologica*, 26, 107–29.

Irish College of Psychiatrists (2005). 'A better future now: position statement on psychiatric services for children and adolescents in Ireland'. Retrieved from www.rcpsych.ac.uk/files/pdfversion/op60.pdf (accessed 12 July 2007).

Irish Sports Council, Lifelong Involvement in Sport and Physical Activity, www.nctc.ul.ie/files/The%20LIPSA%20Model.pdf

Izard, C. E., Fantauzzo, C. A., Castle, J. M., Haynes, O. M., Rayias, M. F. and Putna, P. H. (1995). 'The ontogeny and significance of infants' facial expressions in the first nine months of life', *Developmental Psychology*, 31, 997–1013.

James, W. (1890). *The Principles of Psychology*, Dover Publications.

Kagan, J., Reznick, J. S. and Snidman, N. (1988). 'Biological basis of childhood shyness', *Science*, vol. 240, issue 4849, 167–71.

Kastenbaum, R. (2000). *The Psychology of Death* (3rd edn). New York: Springer Publishers.

Kingston, L. and Prior, M. (1995). 'The development of patterns of stable, transient and school-age aggressive behavior in young children', *Journal of American Academy of Child and Adolescent Psychiatry*, vol. 34, 348–58.

Kohlberg, L. (1976). 'Moral stages and moralisation: The cognitive developmental approach', in T. Lickona (ed.) *Moral Development and Behaviour: Theory, Research and Social Issues* (pp 31–53). New York: Holt.

Kübler-Ross, E. (2005). *On Death and Dying*. London: Routledge.

Kuhn, D., Nash, S. C. and Brucken, L. (1978). 'Sex role concepts of two- and three-year-olds', *Child Development*, 49, 445–51.

Levinson, D. (1986). 'A conception of adult development', *American Psychologist*, 41(1), 3–13.

McAvoy, H., Sturley, J., Burke, S. and Balanda, K. (2006). 'Unequal at birth: inequalities in the occurrence of low birthweight babies in Ireland', Institute of Public Health in Ireland. Retrieved from www.publichealth.ie/index.asp?locID=489&docID=689 (accessed 2 August 2007).

MacFarlane, A. J. (1975). 'Olfaction in the development of social preferences in the human neonate', *Ciba Foundation Symposium*, 33, 103–17.

McHugh, O. (2006) *Celtic Cubs: Inside the Mind of the Irish Teenager*. Dublin: Liberties Press.

McKeown, K., Pratschke, J. and Haase, T. (2003). *Family Well-being: What Makes a Difference?* Dublin: Department of Social and Family Affairs.

Magnusson, D., Stattin, H. and Allen, V. L. (1985). 'Biological maturation and social development: a longitudinal study of some adjustment processes from mid adolescence to adulthood', *Journal of Youth and Adolescence*, 14,(4) 267–83.

Malina, R. M., Bouchard, C. and Bar-Or, O. (2004). *Growth, Maturation and Physical Activity* (2nd edn). Human Kinetics Europe.

Marcia, J. (1980). 'Identity in adolesence', in J. Adelson (ed.), *Handbook of Adolescent Psychology*. New York: Wiley.

Meggitt, C. (2006). *Child Development: An Illustrated Guide* (2nd edn). Oxford: Heinemann Educational.

Meltzoff, A. (1988). 'Infant imitation and memory: nine month olds in immediate and deferred tasks', *Child Development*, 59, 217–25.

Meltzoff, A. and Moore, M. (1977). 'Imitation of facial and manual gestures by human neonates', *Science*, 198, 75–8.

Miller, J. G., Bersoff, D. M. and Harwood, R. L. (1990). 'Perceptions of social responsibilities in India and in the United States: Moral imperatives or personal decisions?' *Journal of Personality and Social Psychology*, 58, 33–47.

Murphy, L. (1937). *Social Behavior and Child Personality*. New York: Columbia University Press.

Murray Parkes, C. (1998). *Bereavement*. London: Penguin.

National Association for Sport and Physical Education (NASPE) (2005). *Physical Education for Lifelong Fitness* (2nd edn). Human Kinetics.

National Coaching and Training Centre (2003). *Building Pathways in Irish Sport*, (consultation paper), University of Limerick.

National Documentation Centre on Drug Use (2006). *National Report*. Dublin.

Nic Gabhainn, S. and Sixsmith, J. (2005). 'Children's understanding of well-being'. Retrieved from: www.omc.gov.ie/documents/research/ChildrenUnderstandingofWell Being.pdf (accessed 5 May 2007).

Nixon, L., Greene, S. and Hogan, D. (2006). 'Concepts of family among children and young people in Ireland', *Irish Journal of Psychology*, 27(1-2), 79–87.

O'Brien, M., Alldred, P. and Jones, D. (1996). 'Children's constructions of family and kinship', in J. Brannen and M. O'Brien (eds), *Children in Families: Research and Policy* (pp 84–100). London: Falmer Press.

O'Connor, T. G., Rutter, M., Beckett, C., Keaveney, L., Kreppner, J. and the ERA study team (2000). 'The effects of global severe privation on cognitive competence: extension and longitudinal follow-up', *Child Development*, 71, 376–90.

O'Connor, T., Rutter, M. and the ERA Study Team (2000). 'Attachment disorder behavior following early severe deprivation: extension and longitudinal follow-up', *Journal of the American Academy of Child and Adolescent Psychiatry*, 39(6), 703–12.

Olweus, D. (1980). 'Familial and temperamental determinants of aggressive behaviour in adolescent boys: a causal analysis', in H. R. Schaffer, *Social Development* (1996). Oxford: Blackwell.

Papalia, D. E., Olds, S. W. and Feldman, R. D. (2005). *Human Development* (10th edn). New York: McGraw-Hill.

Passer, M. and Smith, R. (2001). *Psychology: Frontiers and Applications*. New York: McGraw-Hill.

Pipher, M. (1994). *Reviving Ophelia – Saving the Selves of Adolescent Girls*. New York: Ballantine Books.

Radke-Yarrow, M., Cummings, E. M., Kuczinsky, L. and Chapman, M. (1985). 'Patterns of attachment in two- and three-year-olds in normal families and families with parental depression', *Child Development*, 56, 884–93.

Reitox National Focal Point (2006). National Report to the EMCDDA, Ireland: 'New developments, trends and in-depth information on selected topics'. Retrieved from www.ndc.hrb.ie/toc.php?id=14 (Accessed 10 August 2007).

Rheingold, H. L., Hay, D. F. and West, M. J. (1976). 'Sharing in the second year of life', *Child Development*, 47, 1148–58.

Rich, S. (2005). 'FAS: preventable tragedy', *Psychiatric News*, vol. 40(9), 12. Retrieved from: http://pn.psychiatryonline.org/cgi/content/full/40/9/12 (accessed 25 August 2007).

Rochat, P. (2001). *The Infant's World*. Cambridge, MA: Harvard University Press.

Rosenberg, M. (1985). 'Self-concept and psychological well-being in adolescence', in R. L. Leahy (ed.), *The Development of the Self* (pp. 205–46). Orlando, FL: The Academic Press.

Royal College of Psychiatrists (2004). Factsheet 2: 'Good parenting: for parents and teachers', *Mental Health and Growing Up* (3rd edn). Retrieved from www.rcpsych.ac.uk/mentalhealthinformation/mentalhealthandgrowingup/2goodparenting.aspx (accessed 10 June 2007).

Rushton, J. (2004). 'Genetic and environmental contributions to pro-social attitudes: a twin study of social responsibility'. Retrieved from: http://psychology.uwo.ca/faculty/rushtonpdfs/RoyalSociety.pdf (Accessed 21 August 2007).

Rushton, J. P., Fulker, D. W., Neale, M. C., Nias, D. K. B. and Eysenck, H. J. (1986). 'Altruism and aggression: the heritability of individual differences', *Journal of Personality and Social Psychology*, 50, 1192–8.

Rutter, M. (1989). 'Pathway from childhood to adult life', *Journal of Child Psychology and Psychiatry*, 30, 23–51.

Rutter, M. and Bartak, L. (1973). 'Special education treatment of autistic children: a comparative study: I. Follow-up findings and implications for services', *Journal of Child Psychology and Psychiatry*, 14, 241–70.

Rutter, M. and the English and Romanian Adoptees (ERA) Study Team (1998). 'Developmental catch-up, and deficit, following adoption after severe global early privation', *Journal of Child Psychology and Psychiatry* 39(4), 465–76.

Ryan, S. and Ní Chionnaith, M. (2005). 'Fetal alcohol spectrum disorders: students and school'. Retrieved from www.tcd.ie/niid/research/FASD.pdf (accessed 20 July 2007).

Sacks, O. (1987). *The Man who Mistook his Wife for a Hat*.

Santrock, J. W. (2006). *Lifespan Development*, (10th edn). New York: McGraw-Hill.

Schaffer, H. R. (1996). *Social Development*. Oxford: Blackwell.

Schaie, K. W. and Willis, S. L. (2000). 'A stage theory model of adult cognitive development revisited', in B. Rubinstein, M. Moss and M. Kleban (eds), *The Many Dimensions of Aging: Essays in Honor of M. Powell Lawton* (pp. 175–93). New York: Springer.

Schlundt, D. G. and Johnson, W. G. (1990). *Eating Disorders, Assessment and Treatment*. Boston: Allyn and Bacon.

Schuck, S. Z. *et al.* (1971). 'Sex differences in aggressive behaviour subsequent to listening to a radio broadcast of violence', *Psychological Reports*, 28 (3), 931–6.

Shaffer, D. R. (1999). *Developmental Psychology: Childhood and Adolescence* (4th edn). Pacific Grove, CA: Brooks/Cole Publishing Company.

Simner, M. L. (1971). 'Newborns' response to the cry of another infant', *Developmental Psychology*, 5, 136–50.

Skuse, D. (1984a). 'Extreme deprivation in early childhood – I. Diverse outcomes for three siblings from an extraordinary family', *Journal of Child Psychology and Psychiatry*, 25(4), 525–41.

—. (1984b). 'Extreme deprivation in early childhood – II. Theoretical issues and a comparative review', *Journal of Child Psychology and Psychiatry*, 25(4).

—. (1985). 'Technical note: extreme deprivation in early childhood: a reply', *Journal of Child Psychology and Psychiatry*, 26(5), 827–8.

Smith, P., Cowie, H. and Blades, M. (2003). *Understanding Children's Development* (4th edn). London: Blackwell.

Solomon, D., Watson, M., Delucchi, K. L., Schaps, E. and Battistich, V. (1988). 'Enhancing children's prosocial behavior in the classroom', *American Educational Research Journal*, 25(4), 527–54.

Sternberg, R. J. (1988). *The Triarchic Mind: A New Theory of Human Intelligence*. New York: Viking.

Stratton, K., Howe, C., Battaglia. F. (eds) (1996). *Fetal Alcohol Syndrome: Diagnosis, Epidemiology, Prevention, and Treatment*. Washington DC: National Academy Press.

Thomas, A. and Chess, S. (1977). *Temperament and development*. New York: Brenner/Mazel.

—. (1986). 'The New York longitudinal study: from infancy to early adult life', in R. Plomin and J. Dunn (eds), *The Study of Temperament: Changes, Continuities and Challenges*. Hillsdale, NJ: Erlbaum.

Trivers, R. L. (1971). 'The evolution of reciprocal altruism', *Quarterly Review of Biology*, 46, 35–57.

Tully, M. A., Cupples, M. E., Chan, W. S., McGlade, K. and Young, I. S. (2005). 'Brisk walking, fitness, and cardiovascular risk: A randomized controlled trial in primary care', *Preventive Medicine*, 41(2), 622–8.

UNICEF (2004). *Low Birthweight: Country, Regional and Global Estimates*. New York.

Weber, J. A. and McCormick, P. (1992). 'Alateen members' and non-members' understanding of alcoholism', *Journal of Alcohol and Drug Education*, 37(3), 74–84.

Western Health Board (1998). *Accidents*. Retrieved from www.whb.ie/PublicHealth/file,208,en.PDF (accessed 11 July 2007).

Whiting, B. and J. (1975), *Children of Six Cultures*. Cambridge, MA: Harvard University Press.

Young-Hyman, D., Tanofsky-Kraff, M., Yanovski, S. Z., Keil, M., Cohen, M. L., Peyrot, M. and Yanovski, J. A. (2006). 'Psychological status and weight-related distress in overweight or at-risk-for-overweight children', *Obesity*, 14, 2249–58.

Zahn-Waxler, C., Radke-Yarrow, M. and King, R. A. (1979). 'Child-rearing and children's prosocial initiations toward victims of distress', *Child Development*, 50, 319–30.

Zahn-Waxler, C., Radke-Yarrow, M., Wagner, E. and Chapman, M. (1992). 'Development of concern for others', *Developmental Psychology*, 28, 126–36.

INDEX